Martin Library, York PA

D0872584

Praise for
Last Harvest

"From the initial planning to the home construction to the selling of properties, the five-year project was a challenge for everyone involved, and the author was there every step of the way. Woven into his story are reflections on how American architectural tastes have changed throughout history and how we balance our individuality with [the] need for community."

—Jennifer Caesar, *New York Post*

"Understanding the business of real estate requires an appreciation of its many aesthetic, economic, historical, physical, political, psychological, structural, and countless other aspects, and *Last Harvest* is a primer on them all."

—Henry Petroski, author of *Success through Failure: The Paradox of Design*, writer for *The New York Sun*

"Entertaining and frequently enlightening."

—Penelope Green, *The New York Times Book Review*

"Compelling . . . wonderfully readable . . . Rybczynski is the most fair-minded of writers and absolutely immune to the seductions of current wisdom."

—Andrew Ferguson, *The Wall Street Journal*

"Terrific . . . Steeped in historical knowledge."

—Russ Juskalian, *USA Today*

"Rybczynski is a graceful, personable writer whose considerable erudition is in service to his storyteller's curiosity."

—Lloyd Rose, *The Washington Post*

"Rybczynski has written about the world of American architecture with a simple, rare clarity."

—Annie Dawid, *The Oregonian*

"Rybczynski has a prose style so well designed, even dry bits of design history are as absorbing as a down-filled sofa."

—David Colman, *The New York Times*

"Rybczynski provides historical and cultural perspectives in a style reminiscent of Malcolm Gladwell, debunking the myth of urban sprawl and explaining American homeowners' preference for single-family dwellings."

—*Publishers Weekly*

"A smart, swift tour through the country's embrace of ranch-style and split-level homes in decades past."

—Daniel McGinn, Newsweek.com

"In *Last Harvest,* acclaimed architecture writer Witold Rybczynski paints a nuanced portrait of the men who built America, neighborhood by neighborhood . . . Rybczynski offers a rich history of U.S. development, from the Founding Fathers to the dreamers who conceived the first automobile suburbs in the early part of the last century and on to today's foot soldiers in the movement called New Urbanism."

—David Rocks, *BusinessWeek*

"[Rybczynski is] a perpetually curious observer of architecture and urbanism."

—Teresa Weaver, *The Atlanta Journal-Constitution*

"Maybe you like the way America is being built, maybe you don't, but either way you will not find a more absorbing or patient look at the *real* real estate process than this elegant time-lapsed portrait of a neighborhood to be. Witold Rybczynski is the poet laureate of what you haven't noticed that's probably right in front of you."

— Robert Sullivan, author of *Cross Country: Fifteen Years and 90,000 Miles on the Roads and Interstates of America*

"Nowhere do pretty hypotheses get blast-tested by the facts as in the work of Witold Rybczynski. He is not the kind of scholar who looks at perfectly functional realities and asks whether they can possibly work in theory. Instead, in *Last Harvest* Rybczynski is our engaging and authentic guide, immersing us in a fascinating narrative of how real people live, work, play — and build. *Last Harvest* is *The Soul of a New Machine* for the new urbanism."

— Joel Garreau, author of *Edge City: Life on the New Frontier*

"The master has done it again. His best book ever. It is a must-read for anyone interested in urban development. This seemingly simple example sheds light on suburban development and, in the process, why affordable housing is disappearing."

— Dr. Peter Linneman, Albert Sussman Professor of Real Estate, professor of finance, business, and public policy, the Wharton School, University of Pennsylvania

OTHER BOOKS BY WITOLD RYBCZYNSKI

Paper Heroes
Taming the Tiger
Home
The Most Beautiful House in the World
Waiting for the Weekend
Looking Around
A Place for Art
City Life
A Clearing in the Distance
One Good Turn
The Look of Architecture
The Perfect House
Vizcaya (with Laurie Olin)

LAST HARVEST

HOW A CORNFIELD BECAME NEW DALEVILLE:
Real Estate Development in America from George Washington
to the Builders of the Twenty-first Century,
and Why We Live in Houses Anyway

WITOLD RYBCZYNSKI

SCRIBNER
New York   London   Toronto   Sydney

4927521

SCRIBNER
A Division of Simon & Schuster, Inc.
1230 Avenue of the Americas
New York, NY 10020

Copyright © 2007 by Witold Rybczynski

All rights reserved, including the right to reproduce this book or portions thereof in
any form whatsoever. For information address Scribner Subsidiary Rights Department,
1230 Avenue of the Americas, New York, NY 10020

First Scribner trade paperback edition May 2008

SCRIBNER and design are trademarks of The Gale Group, Inc. used under license
by Simon & Schuster, Inc., the publisher of this work.

For information about special discounts for bulk purchases, please contact
Simon & Schuster Special Sales at 1-800-456-6798 or business@simonandschuster.com

DESIGNED BY ERICH HOBBING

Text set in Stempel Garamond

Manufactured in the United States of America

1   3   5   7   9   10   8   6   4   2

Library of Congress Control Number: 2006052136

ISBN-13: 978- 0-7432-3596-9
ISBN-10:   0-7432-3596-7
ISBN-13: 978- 0-7432-3597-6 (pbk)
ISBN-10:   0-7432-3597-5 (pbk)

Photograph, page 1: courtesy of Arcadia Land Company
Illustration, page 90: courtesy of Robert R. Heuser
Photographs, pages 9, 185, 275, and 283: courtesy of the author

To Shirley

God made the country, and man made the town.
—WILLIAM COWPER

CONTENTS

CONTENTS

PART ONE

New Daleville, September 2003

Prologue

Twenty years ago, my wife and I started to walk for exercise, every morning before breakfast. We lived in the country, and our route was a winding road between meadows and apple orchards. Since moving to Philadelphia, we walk on city streets. The experience is different, yet not so different.

Chestnut Hill, where we live, is as bucolic as its name. There is a hill, and there are horse chestnut trees, though the American chestnuts that gave the place its name are long gone. Our walks take us down arboreal tunnels of massive oaks and sycamores, which grow in wide planting strips between sturdy granite curbs and slate sidewalks. The strips, which are the responsibility of individual homeowners, exhibit a pleasant disharmony. Most people, following an unwritten rule, plant grass, but there are also nonconformist patches of ground cover, defiantly individualistic flower beds, no-nonsense brick pavers, mean-spirited bands of crushed stone, and in at least one case, an earnest row of zucchini.

The boundaries of the house lots are likewise variously defined. Many are generously open; some have hedges or planting beds. There are ivy-covered wooden fences of every sort, as well as black wrought-iron railings, white pickets, and the occasional stone wall. A few houses have solid wooden fences, unsociable barriers that resemble stockades out of *The Last of the Mohicans*.

Houses change with the seasons. Pots of flowers appear on

stoops, and wreaths adorn front doors. The decorations on our neighbor's lawn are always a treat: ghosts for Halloween, angels at Christmastime, pink flamingos for the children's birthdays. Last Valentine's Day, every window contained an illuminated heart. Some houses fly flags. Not as many Stars and Stripes as immediately after 9/11, but several of those odd flower-power banners that people seem to like. Dave, an ex-Marine, hoists the red standard of the Corps. We sometimes meet him in the morning, watering the rosebushes in front of his house. Most Chestnut Hill houses are close to the sidewalk. So close you can look inside.

Garbage day is a sort of public confessional. You can see who's bought a new computer, and who's given up on the exercise machine. The other morning I came across a discarded tabletop hockey game. For a second, I thought of lugging it home. When the contents of basements and attics appear on the sidewalk, it means a move is imminent. Families come and go with regularity; finally, we're all of us just passing through. A young household moves in, and swing sets sprout in the backyard. If the children are older, it's a basketball hoop. A new owner usually means energetic gardening, at least for a season or two. Gardens are the main things that change. Occasionally, someone adds a terrace to a house, or encloses a porch. New owners undertake long-delayed maintenance: putting on a new roof, or repointing walls. The only significant construction on my street in several years has been the repair to a neighbor's house that was hit by a falling tree. When the work was finished, the house looked exactly as it did before the accident. After all, why improve on a good thing?

There is no typical Chestnut Hill house. There are mansions as big as small hotels, and little Hansel-and-Gretel cottages. Our walks take us by a representative sample of the architectural styles that came and went during the late-nineteenth and early-twentieth centuries: charming Queen Annes with picturesque bay windows and ornamental curlicues; rather serious, half-timbered Tudors; elegant Georgian Revivals that make me think of

Jazz Age financiers in wing collars and spats; and straightforward center-hall Colonials, as friendly and uncomplicated as the big golden Labs that play in their front yards. One street has a row of flinty stone cottages that appear to have been transported directly from the Cotswolds. Schist, quarried from a nearby ravine, is the common building material, but we also see brick, stucco, and clapboard. If we walked farther than our usual three miles, we would pass Italianate, Jacobean, and Romanesque Revival residences. Not all the houses are old. Beginning in the nineteen fifties, some of the large estates were subdivided. The grandest of these properties was Whitemarsh Hall, a celebrated Gilded Age mansion, designed by Horace Trumbauer in 1917 for Edward T. Stotesbury. All that's left of the 145-room Georgian pile is a pair of huge entrance gates, whose massive columns loom over the plain-Jane bungalows that dot the grounds of what was once a formal French garden.

What drew Stotesbury, a stockbroker and banker who was reputedly the richest man in Philadelphia and owned second homes in Bar Harbor, Maine, and Palm Beach, Florida, here? During most of the nineteenth century, Chestnut Hill had been a sleepy rural hamlet. Summer visitors included Edgar Allan Poe and John Greenleaf Whittier, who came to experience the rugged landscape of nearby Wissahickon Creek, and wealthy Philadelphians, who, attracted by the salubrious climate, built country estates. In 1854, thanks to a consolidation of city and county, Chestnut Hill became part of Philadelphia, but remained largely rural. In the eighteen eighties, Henry Howard Houston, a wealthy local businessman, bought up 3,000 acres of this countryside, and after convincing the Pennsylvania Railroad, of which he was—not by chance—a director, to build a commuter line from downtown to Chestnut Hill, he set about subdividing the land and developing a new community. He called it Wissahickon Heights. To give his development social cachet and attract Philadelphia's elite, he founded the Philadelphia Horse Show, which became the premier social event of the city. To draw summer visitors, he built a large hotel—complete with an artificial lake. He added a country club, where residents could drink and play cricket (a popular game in Anglophile Philadelphia, which had several cricket clubs) and built a picturesque Gothic church, where they could worship their Episcopal God.[1]

Garden suburbs such as Wissahickon Heights were part of an important episode in American urban history, when upper-middle-class families moved from the centers of cities to their suburban fringes. The Harvard historian John Stilgoe has called these outlying communities "borderlands." He reminds us that this displacement was the mark of cultural as well as physical transformation. "The enduring power of borderland landscape between the early nineteenth century and the beginning of World War II," he writes, "suggests that many women and men understood more by *commuting* and *country* than train schedules and pastures, and hints also that the cities of the Republic failed to provide an urban fabric as joyous, as restorative as that found by borderers a few miles beyond."[2]

Borderers were not back-to-the-landers. They expected attractive, urbane residences, cultivated landscapes—and cultivated neighbors—which required organization. A few of the nineteenth-century borderland communities grew spontaneously, but most, like Wissahickon Heights, were planned. The first, by most accounts, was Llewellyn Park in New Jersey, developed in 1853 by Llewellyn Haskell, a Manhattan businessman, and designed by the celebrated architect Alexander Jackson Davis. The largest was Riverside, Illinois, laid out by Frederick Law Olmsted and Calvert Vaux in 1868 for the Chicago developer E. E. Childs. Similar communities appeared on the outskirts of every major American city.

So many Philadelphians found Houston's development "joyous and restorative" that, by the time Stotesbury moved here, Chestnut Hill was the city's most prestigious address. Houston's son-in-law, George Woodward, greatly expanded the business during the early nineteen hundreds.* A physician with an entrepreneurial streak, he was also a progressive philanthropist, interested in architecture and social housing. He subdivided land and sold lots to wealthy Philadelphians, but he also built a variety of rental houses—middle-class family homes as well as large residences, a range that continues to give the neighborhood a diverse charm. Woodward hired young architects whom he'd send to England to broaden their repertoire—and to discover Cotswold cottages. In 1921 he commissioned Frederick Law Olmsted, Jr., son of the famous landscape architect, to lay out a public park. Woodward's descendants continue to manage rental properties in Chestnut Hill to this day.

My own house—built on the foundation of an old icehouse— was designed for Woodward's development in a Colonial Revival style by H. Louis Duhring in 1908. Stepped gables give it a Dutch appearance, and the interiors are rustic, with pegged, rough-hewn beams and fieldstone fireplaces. The public rooms, following the British Free Style, are exceptionally open. It's a testimony to Duhring's talent that the house has served as a family home for almost a hundred years with only minor modifications. The bedrooms were remodeled in the nineteen thirties, extra bathrooms added in the fifties, a porch turned into a sunroom in the sixties, and the kitchen renovated in the nineties. I converted two bedrooms into a study when we moved here six years ago. No doubt, during the twenty-first century it will undergo more alterations. If energy costs continue to rise, there will come a time—I hope it's not on my shift—when someone will have to figure out how to add proper insulation to the walls. But the original roof slates have lasted, and the roofer assures me that, with a little care, they're good for some time yet. The stone walls need periodic repointing, and the woodwork must be properly maintained. All of us own-

---

*Woodward changed the name of the development to St. Martin's, which survives as the name of a train stop.

ers over the years have performed these essential tasks, driven by the house's simple but sturdy details, its practical plan, and its intrinsic good character.

Not far from my home is an unusual group of houses that Duhring built for Woodward in 1931. By that time, many people owned cars, and Roanoke Court, as it's called, is entered through a walled motor court flanked by individual garages that resemble two rows of stables. Beyond that, eight attached houses surround a common garden. It's a magical, secluded space. The large houses are designed in a simplified version of the English vernacular style that Duhring favored, with steep slate roofs and rough stone walls. He built a number of such novel housing groups in Chestnut Hill, including several courts, a crescent of semidetached residences, and a cluster of unusual quadruple houses. Woodward encouraged such experimentation. In addition to the Cotswold row, he commissioned a lane of British country-style cottages and a cluster of charming Norman houses, complete with a town gate, known locally as French Village. The last was built after the First World War to honor Woodward's deceased son, a pilot in the Lafayette Escadrille.

Visitors to Chestnut Hill use terms such as *old-fashioned* and *traditional* to describe the treed streets and interesting-looking houses. They can be forgiven for assuming that the neighborhood is the result of years of fortuitous evolution—a suburban version of Nantucket or Martha's Vineyard. Nothing could be further from the truth. Evolution there has been, but pastoral Chestnut Hill is no happy accident. It was a residential real estate development, and it was *designed* to look the way it does.

1

The Developer

"The construction side is almost risk-free, since building begins only after the house has been sold to a buyer. All the risk is in the development side, but so is all the money. A small home builder makes 5 to 7 percent profit, while a developer can make a lot more than that—or he can go bankrupt."

Every spring I invite Joe Duckworth, a residential developer, to talk to my class, a mixture of architects, planners, and Wharton School MBAs. He generally begins by reminding the students that home building is an unusual business. "The customers are not only buying a product," he says. "They're looking for the right location for commuting to work, good schools, recreational amenities, and nice surroundings. They're shopping for a neighborhood."

He shows images of suburban communities, asking the students to describe what they see. "Lawns," they answer. "Colonial shutters." "Brick chimneys." Emboldened, someone in the back calls out, "Boring, cookie-cutter houses." "Interesting answer," says Duckworth. "You're right, the houses are similar. When people buy a house, they want to be able to sell it. Since they can't afford to lose money, they're highly risk-averse. They want what everyone else has."

Paul, an architecture student, raises his hand. "The houses that you're showing all look pretty traditional. What's the market for modern design?" Duckworth answers that in the seventies a home builder he worked for created a so-called California contemporary model, with clerestories, cedar siding, high spaces, and an open plan. "It wasn't great architecture, but it was different. Today, those houses are selling at a ten to twenty percent discount compared to other nineteen-seventies-era houses. People just don't like them, and no Philadelphia builder has tried it since."

Duckworth talks about his business. "In the past, residential development was straightforward," he says. "You had an engineer prepare a subdivision plan, you got it approved, and built the houses. Development and building were done by the same company. In the last five years, thirty-eight states have enacted some kind of land development regulations. Today, especially in an anti-growth area such as Pennsylvania, getting land permitted is an art that requires a different skill set than building houses, so land development and house building are increasingly done by different people. Development involves acquiring land, getting permits, and putting in roads and infrastructure; house building is mainly about construction. The construction side is almost risk-free, since building begins only after the house has been sold to a buyer. All the risk is in the development side, but so is all the money. A small home builder makes five to seven percent profit, while a developer can make a lot more than that—or he can go bankrupt."

Kelly, one of the Wharton students, asks how developers weather economic downturns. "It's mostly a question of resources," Duckworth says. "In a downturn, about a quarter of developers go bankrupt. They've bought land which they can't sell. So the rest of us have the opportunity to buy this land at a low price. When the economy turns up, we have permitted land ready to go, while other developers are just starting the long permitting process."

Duckworth discusses the role of regulation in development.

"You have to understand that the way that our suburbs are planned is not because of developers, it's mainly because of zoning," he tells the class. "Who do you think controls zoning?" he asks. "Zoning boards," calls out a smart aleck. "Yes, but zoning boards are run by who? The local residents. What these people want is to maintain, or even increase, property values. At the same time, they want—and their neighbors want—to limit development as much as possible. In Chester County, where I live, the size of an average lot increased from half an acre in the sixties to one acre in the eighties, and by the end of the nineties it was an acre and a half. The bias of local zoning is always towards bigger lots." Duckworth ends by talking about his own projects. "I'm working on village-type developments with smaller lots and more open space. It's taking a long time to get approvals, though, because we're swimming against the current."

Duckworth and I have been friends for more than a decade, and we usually have lunch after the class. He's in his early fifties, with longish hair and a beard that he's recently been growing and shaving off with disconcerting regularity. Today he's bearded. I tell him that I'm sure the students appreciated his comments since many of them want to be real estate developers. I ask him what attracted him to the field. "I studied mechanical engineering at Carnegie Mellon in Pittsburgh," he says, "and after graduating I got a job with Sun Oil in Philadelphia. After a few months I realized that my future was not in engineering, and I decided to get an MBA and go into business. Like most of my Wharton classmates, I wanted to be an entrepreneur and run my own company. In most fields, that meant spending years working your way up the corporate ladder and then, if you were lucky, having one shot at being CEO. I didn't want that. I was already married with kids, and I was in a hurry. I looked around at business sectors where someone like me, with a college education and an MBA, had an advantage. I came across commercial home building, which I didn't know anything about. It was a field that seemed to have many family-run businesses. I thought that I could bring modern business practices to bear and make my way."

Eventually, he landed a job with Toll Brothers, the largest home builder in the Philadelphia area. His responsibility as assistant to the president was finding and buying land and getting approvals. He learned the business but after nine years left the company. "I realized that I was never going to be a brother," he jokes. He moved to Realen Homes, one of Toll Brothers' smaller competitors. "Realen was a reputable company that owned apartment buildings that generated good cash flow, but the home-building side of the business was not doing well. It had lost money on a deal that went sour, the employees were demoralized, and there were no projects in the pipeline." Duckworth was brought in as president and CEO to revive the operation. Using his Toll contacts, and Realen's credibility as a company, he immediately optioned more than three thousand lots. "Over the next decade I built up company sales from twenty million dollars a year to a hundred million, making Realen the second-largest home builder in the Philadelphia area, after Toll," he tells me.

I know that Duckworth has recently left Realen to start his own real estate company, and I ask him about the projects he mentioned in class. "We're in the middle of trying to get several of these village-type developments off the ground, which requires townships to change their zoning to allow smaller lots. It's an uphill battle," he says. "There is one project that looks promising, though. It just came to me through another developer, Dick Dilsheimer. I've known Dick a long time. He and his brother are old-fashioned merchant builders, that is, they buy land, subdivide it, build reasonably priced houses, and market them to buyers. For the last year they've been trying to get permission to build a small subdivision in southern Chester County. It's nothing special: eighty-six houses on ninety acres of rural land. Dick's problem is that the township doesn't like his project. They keep telling him that they want something different, with smaller lots and more open space."

Duckworth calls Dilsheimer's proposed development "as of right," that is, it follows local zoning exactly and does not require a variance, or special approval. Nevertheless, the township is blocking him. "He could sue and probably win, but confrontation is not Dick's style. Instead, he's approached me to see if I would be willing to take the project off his hands. I'm interested, but it's still too early to know how serious the township really is."

I've heard architects and city planners argue for more density and open space, but here the demand is coming from the citizens themselves. I ask Duckworth if he knows what has pushed the township in this direction. "I'm not sure," he says. "It may have been their planning consultant, Tom Comitta."

I happen to know Tom. We share an interest in garden suburbs. He introduced me to Yorkship Village in Camden, New Jersey, which was built during World War I to house shipyard workers, and I repaid him by showing him Roanoke Court and some of the other residential groups in Chestnut Hill. If he's involved, that might explain a lot. I decide to pay him a visit and find out more about this unusual township.

Tom Comitta lives and works in the town of West Chester, the seat of Chester County. His office occupies half of a brick Victorian twin on Chestnut Street. The sign on the door says, THOMAS COMITTA ASSOCIATES, TOWN PLANNERS & LANDSCAPE ARCHITECTS. Though he is trained as a landscape architect, much of Comitta's business is advising small rural municipalities that need professional help with planning, zoning, transportation, and other development issues. One of his clients is Londonderry Township, the site of Dilsheimer's proposed subdivision, which Comitta calls the Wrigley tract. "Londonderry is a small rural township in southern Chester County, at the edge of the Brandywine Valley," he tells me. "They originally hired me to advise them on a large subdivision of three hundred town homes called Honeycroft Village. It's a nice name, but it was an unimaginative plan with identical houses in groups of threes and fours." The township supervisors were dissatisfied with the layout. "What else can we do?" they asked Comitta. He suggested a visit to a new planned community in another part of the county that would give them an idea of an alternative approach to residential planning.

Considering it was a Saturday morning, the turnout was surprisingly good, he told me. The group included the three township supervisors, the township engineer, members of the planning commission, and a representative of the developer. The new community consisted of large houses, two-car garages, front lawns, and attractive landscaping. But as the group walked around and Comitta pointed out various features, it became apparent that in many small ways this development was different. To begin with, there were sidewalks shaded by trees growing in planting strips. The lots were smaller, the buildings closer together—and closer to the sidewalk. Cars were parked on the streets, but there weren't any driveways or garages—these were in the back, accessed from rear lanes. Many of the houses had front porches and picket fences. These features gave the development a compact, villagelike appearance.

They met a woman driving her car out of a lane. "She'd been living there about eighteen months, and she was rhapsodic," Comitta remembers. "She said that it reminded her of her mother's hometown." The township engineer expressed some skepticism about the narrowness of the streets, but Comitta saw that most of the group were favorably impressed. The visit lasted about two hours. Afterward, they stood around talking. The representative of the Honeycroft developer said he was concerned about the time it would take his client to redesign the plan and go through an entirely new approvals process. Then the chairman of the planning commission said, "We might not be able to do this in Honeycroft, but wouldn't this kind of thing be better for the Wrigley tract?"

"That's how it began," Comitta told me. "The township was unhappy with Dilsheimer's proposal, and the visit suggested an alternative at just the right time. The planning commission approved Honeycroft, but they want Dilsheimer to change his project. They've really given him a hard time, so I can understand why he wants to pull out. Now the township has asked me to work with Joe Duckworth on the Wrigley tract and see if we can get something that will be better than what we've done before."

2

Seaside

How a little resort community with porches and picket fences
became a touchstone for suburban planning.

T he villagelike community that Tom Comitta had shown the
Londonderry Township officials is an example of what is
often called neotraditional development, a planning movement
that began in the nineteen eighties. It was sparked by two events.
In 1981 the Cooper-Hewitt Museum in New York City mounted
an exhibition called "Suburbs." A large part of the show was
historical and featured many of the early garden suburbs, such as
Chestnut Hill and Yorkship Village, which Comitta had shown
me. It also included forgotten classics such as Tuxedo Park outside New York City, Forest Hills Gardens in Queens, and Palos
Verdes Estates in Los Angeles.

The organizer of the exhibition was a forty-two-year-old architect and Columbia University professor, Robert A. M. Stern.
Stern had become interested in the early garden suburbs thanks in
part to Chestnut Hill. "I can distinctly remember Bob Venturi
touring me past French Village in the late sixties," he told me. The
Cooper-Hewitt exhibition made an important polemical point:
suburbs are an integral part of American urbanism. This was a
bold claim. At the time, serious architects considered suburbs and
suburban houses beneath contempt. Not Stern. "The modest

single-family house is the glory of the suburban tradition," he had written earlier. "It offers its inhabitants a comprehensible image of independence and privacy while also accepting the responsibilities of community."[1]

John Massengale, a University of Pennsylvania graduate student working in Stern's office, coedited *The Anglo-American Suburb,* which accompanied the exhibition.[2] It was the first time that many of the developments had appeared in print in over fifty years. What was the inspiration for the book? "Traditional town planning was something that was in the air," Massengale recalls. "There was a general dissatisfaction among young architects with orthodox modernism, especially modernist city planning." One could argue that the unpopularity of modernist houses, which Joe Duckworth had mentioned to my students, was a matter of taste, but there is no question that the modernist city planning policies of the nineteen sixties had been a disaster. Highway construction and urban renewal destroyed neighborhoods, and public housing, though built with the best intentions, by concentrating the poor in high-rise blocks created more problems than it solved.[3]

The architectural reaction to modernism became known as postmodernism. But postmodernism proved too glib and weak-kneed, and unwilling to question the underlying premises of modernism. Tom Wolfe once compared postmodern architecture to Pop Art, calling it "a leg-pull, a mischievous but respectful wink at the orthodoxy of the day."[4] By the nineteen eighties, postmodern architects reached a parting of the ways. Some returned to the fold, so to speak, embracing various modernist revivals: minimalist International Style, early forms of Russian deconstructivism, and sculptural German expressionism. Others, including Stern, sought inspiration in a more distant past. In that sense, the renewed interest in the old garden suburbs should be seen not only as a revival but also as a desire to continue a tradition.

Massengale calls *The Anglo-American Suburb* "the opening salvo in the whole garden suburb renaissance of the eighties." The first fully realized project of that renaissance was not designed by Stern, nor was it even a suburb. Seaside, begun in 1982 and completed over the next two decades, is a holiday resort on the Florida Panhandle, consisting of approximately three hundred houses and roughly the same number of guest cottages, as well as shops, restaurants, and commercial buildings. Most Florida resorts are designed to look like country clubs; Seaside is different. Narrow streets radiate from a central green as in a New England village. The houses are vaguely Victorian, with traditional pitched roofs, porches, and white picket fences. The lots are small and the buildings extremely close together, bordered by heavy undergrowth. Sandy footpaths provide shortcuts behind the gardens. The casual atmosphere and cottagelike houses recall an old-fashioned beach community.

The first time I saw Seaside was in 1989.[5] The place was less than half finished, but it made a powerful impression. I belong to that generation of architects for whom the central issue in architecture is housing. As a student, I dutifully visited the modern housing that was considered exemplary: Le Corbusier's Marseille apartment block, with its famous shopping street in the sky; Mies van der Rohe's Lafayette Park in Detroit, which combined low-rise and high-rise buildings on an urban site; and Louis Kahn's Mill Creek public housing in Philadelphia, then considered a model of its type. Truth to tell, these projects were uniform, standardized, and lifeless. I sensed—even if I didn't quite admit it—that none was as lively as the old Italian and Greek towns and villages I visited on my student trips. I assumed it was just a question of time. Any residential development built all at once was bound to be uniform and somewhat dull, I told myself. Seeing Seaside was a shock, since here was a brand-new development that was neither uniform nor dull; instead it was varied and animated.

Seaside was planned by Andrés Duany and Elizabeth Plater-Zyberk, a young husband-and-wife architect team based in Miami. They had both graduated from Yale in 1974. Duany had briefly worked for Stern, and the couple had contributed to *The Anglo-American Suburb*. Although Duany and Plater-Zyberk were among the cofounders of Arquitectonica, a chic Miami architectural firm, they had since moved away from modernism and become interested in traditional urbanism. They both taught at the University of Miami, where they did town planning projects with students, studying old Florida towns such as Key West. Not coincidentally, they lived in the garden suburb of Coral Gables.

George E. Merrick, who developed Coral Gables in the twenties, grandly called his project "America's treed suburb." Later planned suburban developments were known as "subdivisions," and their developers as "subdividers." Over time, *subdivision* acquired a pejorative connotation and was supplanted by the more wholesome *community,* as in *golf course community* and *retirement community.* But Duany and Plater-Zyberk did not refer to their project as a resort community—which is what it was—they called it a town.

The small town occupies an iconic position in American popular culture. All countries have small towns, of course, but in the United States the small town embodies a particular ideal of neighborly democracy, self-sufficiency, and independence. In the mid-nineteenth century, Ralph Waldo Emerson wrote that "the town is the unit of the Republic," but the popular image of the small town really came into its own a hundred years later.[6] Artists as disparate as Mark Twain, Thornton Wilder, Frank Capra, and Norman Rockwell stoked the small-town myth. So did Walt Disney, who made a small-town main street the centerpiece of his first theme park. Such images penetrated the public consciousness. When a 1990 Gallup poll asked people where they would prefer to live, despite the fact that four out of five of the respondents resided in a metropolitan area, small towns were strongly favored over suburbs, farms, or cities.[7] By calling Seaside a town, planning it like a town, and incorporating small-town features such as picket fences and front porches, Duany and Plater-Zyberk were tapping into a powerful cultural tradition.

Time, which featured Seaside in its 1990 "Best of the Decade" issue, speculated that "the 1990s might be ripe for the Seaside model . . . to become the American planning paradigm." Between 1988 and 1990, Duany and Plater-Zyberk designed two dozen Seaside-like planned communities across the country. Although the recession of 1990 stalled or halted most of these projects, the end of the decade saw several new garden suburbs take shape, some designed by Duany and Plater-Zyberk, and some by others. The largest and best-financed was built by the Walt Disney Company near Orlando, Florida.[8] The new town of Celebration included a high school, a primary school, and a health care facility, as well as a full-fledged town center next to a lake. The first phase of what would eventually house ten thousand was inaugurated in 1996. Closing the circle that had begun fifteen years earlier, one of Celebration's architects and planners was Robert A. M. Stern.

Despite the publicity, this handful of developments was hardly the new paradigm that *Time* foretold—it was a drop in the bucket among the tens of thousands of suburban developments built during that period. Yet the impact of the new generation of garden suburbs has been greater than their small number might suggest. This is thanks largely to Duany and Plater-Zyberk, who in addition to being talented planners are zealous and energetic advocates. They have codified the Seaside approach and coined the term *traditional neighborhood development,* or TND. They created a foundation that distributes information to interested municipalities, and they convinced President Clinton's Department of Housing and Urban Development to incorporate traditional neighborhood principles into its inner-city housing projects. They conduct workshops and courses for the Urban Land Institute, the research and education arm of the real estate industry, which has endorsed traditional neighborhood development as a type of suburban planning. They are also cofounders of the Congress for the New Urbanism, which has become the prime forum for planners and architects interested in the subject. Thanks to the influence of Duany and Plater-Zyberk, new, large neotraditional urban neighborhoods have appeared in Denver, Albuquerque, and Orlando.

Andrés Duany is harshly critical of conventional suburban planning. "The classic suburb is less a community than an agglomeration of houses, shops, and offices connected to one another by cars," he says, "not by the fabric of human life."[9] His point is that suburbs have the right ingredients but that they are improperly put together, strung out along collector roads, functionally segregated, housing over here, office buildings over there, shopping elsewhere. "These elements are the makings of a great cuisine, but they have never been properly combined," he says. "It is as if we were expected to eat, rather than a completed omelet, first the eggs, then the cheese, and then the green peppers."[10]

In a typical lecture he shows a slide of contemporary town houses, stepped back in a sawtooth pattern, with desultory landscaping and parking slots facing the front doors. He contrasts this banal arrangement with a street scene in Old Town Alexandria, Virginia. He points out that the basic elements—attached row houses, asphalt, parked cars—are similar. He talks about how, in Alexandria, façades line up to form a wall defining the street, how slight variations between the houses make all the difference, how the sidewalk and the street trees separate the houses from the cars parked on the street. You don't have to be a town planner to see which one is better. "The market shows that people are willing to pay several times as much to live in Old Town Alexandria as they are to live in a modern townhouse in a typical development, several times as much for termite-ridden beams and parking that on a good day is two blocks away."[11] Duany delivers the punch line with a flourish, like a conjurer pulling a rabbit out of his hat.

3

Epiphanies

Learning that, under the right circumstances, home buyers would not only accept density but actually pay more for it, was the real estate equivalent of discovering a new planet.

Joe Duckworth saw Seaside in 1996, while attending one of Andrés Duany's workshops. Duckworth didn't agree with all that the architect said, and he thought that some of his views about real estate development were more than a little simplistic, but he was impressed by Seaside. He made a point of meeting its developer, Robert Davis.

Seaside is widely known among architects thanks to Duany and Plater-Zyberk's innovative planning, but the beach community would not have achieved its popular fame had it not also been a commercial success. That was largely Davis's doing. The way he built Seaside was surprisingly conservative. He had been a real estate developer in the nineteen seventies in Miami, where a brush with financial failure made him averse to partners and debt. Davis developed Seaside slowly. Initially, he sold only twenty to thirty lots a year—he did no home building himself—and reinvested the profits in infrastructure. He cut overhead to the bone. For example, since the lots were too small for individual septic tanks, the first group of houses was connected to a common septic field located in an unbuilt portion of the site. Only after sell-

25

ing more lots did he install a sewage treatment plant, which he upgraded as the community grew. Davis, who has a Harvard MBA, is a clever businessman. When he was unable to sell some awkward, pie-shaped lots around a small circular plaza, he built a gazebo in the plaza; the lots sold like hotcakes—at a premium. To encourage house sales, he started a rental program, which not only allowed homeowners to recoup part of their costs but also generated steady income. In addition, many renters later became buyers. He carefully nurtured small local businesses and slowly created a lively town center with kiosks, shops, restaurants, and bars, which attracted many day visitors. Like all good developers, he took care of the details.

In 1982 Davis sold the first lot for $15,000, slightly more expensive than lots in nearby beach communities; after four years, he was getting $50,000. That was only the beginning. Thanks to the healthy economies of southern Georgia and Mississippi, real estate on the Gulf was booming. By 1992 he was selling lots for $130,000, and by 2001 for $690,000. Waterfront lots, which he had wisely held on to, were going for close to $2 million.[1] In 1981 Davis's land had been valued at less than a million dollars; by 1996, when Duckworth first visited Seaside, the assessed value of the entire development was approaching $200 million.

"I don't want to be the first one taking a bite of the apple," Duckworth often says, so he appreciated Davis's pioneering accomplishment. The working assumption of residential developers had always been that Americans dislike density. The selling price of a home is usually in direct proportion to the size of the lot, the ideal being a lot so large that you can't see your neighbors— the proverbial house in the woods. That was the image Duckworth sold at Realen. Davis had demonstrated that, under the right circumstances, home buyers would not only accept density but actually pay more for it. This was the real estate equivalent of discovering a new planet.

At the time Duckworth visited Seaside, he was ready to make

a change himself. His career at Realen was a success. He had learned the formulas and rules of thumb of the home-building business, and he was good at it. In 1992 *Professional Builder* magazine named him National Builder of the Year. But the work was no longer a challenge. The truth was that he found it repetitive and uninteresting. "I like the creative side of the business, but home building is only five percent about development, the rest is making and selling a consumer product," he says. He spoke to members of the board, and they suggested that he appoint a CEO. Then, he says, he could "kick [himself] upstairs and join them on the golf course." That didn't appeal to him. "I wasn't ready to retire," he says. "Anyway, I don't play golf."

Duckworth had always been interested in architecture and design, and he had served as head of Philadelphia's Architecture Foundation. But most of his work at Realen was distinctly conventional as far as design was concerned. He had done a couple of what he called "progressive developments" that went beyond the cookie-cutter format, clustering houses on small lots, preserving open space and farmland. One of these projects had won a planning award. "I liked doing this kind of development, but it always took more time, so I couldn't do much of it at Realen," he recounts. "I figured that, if I was on my own, I could take on more projects dealing with issues such as environmental conservation and walkable communities."

Duckworth decided he would start his own company, doing only land development, not home building. "From my Wharton classes I remembered product differentiation and market segmentation," he says. "I figured I would focus on unusual and unconventional real estate opportunities, and special clients, such as old family landowners, institutions, or progressive townships, who were interested in quality and in a long-term investment rather than in quick returns." The first person he told about his planned career change was his friend Chris Leinberger, a member of Realen's board. Leinberger was a managing director at a national real estate consulting firm. His reaction was unexpected. "You know, Joe, I'm bored, too. I think I might join you." Then he added, "You met Robert Davis when you were at Seaside. I think he might be interested."

The trio formed the Arcadia Land Company, a loose partnership in which everyone would have a percentage of every project but one partner would always be in charge. "We complement each other nicely," says Duckworth. "I know land development. Chris, who is based in Santa Fe, is a land-use strategist. Robert is not a big deal maker, but he's the one with the vision and the national reputation. He's the man who built Seaside."

Tom Comitta heard Andrés Duany lecture about Seaside during a Harvard summer program. Comitta has firsthand experience of old-fashioned urbanism,; he grew up in Manayunk, a blue-collar neighborhood of Philadelphia where his father had a barbershop. With its steep streets and narrow houses, Manayunk resembles an Italian hill town, but what drew Calabrian immigrants like Comitta's grandfather here in the twenties were the paper mills lining the Schuylkill River. Today, Manayunk's Main Street is a fashionable restaurant row, but while Comitta was a boy, it was still a neighborhood shopping street.

In the summers, Comitta worked for his maternal grandfather, who tended the grounds of estates on the Main Line, and when he was admitted to Penn State, he majored in landscape architecture. He won a scholarship to Harvard. After graduating, he returned to Pennsylvania and was offered a job by a large environmental engineering firm in West Chester. He worked for the firm's municipal clients, writing floodplain ordinances, preparing environmental impact assessments, and reviewing zoning and development applications. After two years he felt ready to strike out on his own. He married, started a family, settled down.

In 1993 his father became terminally ill. Comitta found himself taking stock of his own life. It wasn't a midlife crisis, exactly, though he was forty-four, but looking back over the last twenty years, he realized that he wasn't satisfied with the direction his career had taken. The small municipalities and rural townships he worked for were under pressure from the suburban growth of Philadelphia and Wilmington, and a large part of his job was helping them evaluate applications for new residential subdivisions, office parks, and strip malls. "The truth was that my clients were unhappy with what they were approving," he says. "They dragged out the process. They beat up on developers. And they didn't like the results when they were built."

Comitta felt he needed to rethink what he was doing. During Duany's workshop, he had learned about Raymond Unwin, an English architect and planner who was a central figure in the British garden suburb movement of the early nineteen hundreds. Unwin's masterpiece was a large residential development outside London, Hampstead Garden Suburb, which Robert A. M. Stern has called "the jewel in the suburban crown."[2] Comitta found a copy of Unwin's long out-of-print primer, *Town Planning in Practice*, in the library.[3] The illustrated text described precise principles for designing new communities.

Comitta's father died in September, and he decided to take a month's leave and go to England to see Unwin's work for himself. He visited Hampstead Garden Suburb and found the experience absorbing. "I would read Unwin's book in the evening, then walk around the actual places he was writing about the next day," he recalls. Unwin had modeled parts of the suburb on Rothenburg, a thirteenth-century Bavarian town, so Comitta, who wanted to see the original, went to Germany.

A planner from suburban Pennsylvania looking for lessons in a medieval town in Germany sounds odd. Surely the two situations are so different as to defy comparison. But town planners — like architects — have to see things with their own eyes. Comitta had spent many years intellectualizing his profession, worrying about rules and regulations. He needed to be reminded of the physical reality of planning. "Walking around these old towns, I saw the delicate relationships that exist between large spaces and small, the progression from narrow to grand as one passes from street to plaza," he explains.

Comitta stretched his leave to almost two months. Shortly after returning home, he invited his largest clients to a slide show at the town hall. It wasn't exactly "My European Vacation." Comitta told the audience that the old ways of mixing uses, such as residences, shops, and community buildings, were more effective than modern zoning, which separates uses into different areas. The point was not to copy English garden cities and German medieval towns, he said, but to look to our own urban traditions, the old small towns of southeastern Pennsylvania. His message was that, rather than resist new development, municipalities should actively direct growth to complement and improve their communities. He spoke in his usual calm and deliberate fashion, but in the context of southeast Pennsylvania—or, indeed, in the context of almost anywhere in the metropolitan United States—Tom Comitta was preaching revolution.

"Some people must have been put off by the pictures of the dense center of old Rothenburg, for the next day, two of the townships called and politely fired me," he says. "On the other hand, two others told me they liked what I had to say." Encouraged, Comitta started writing new kinds of zoning ordinances that would allow small rural communities to grow in ways more compatible with how they had done it for the last hundred years. He says, "I tried to figure out what people liked about their old towns and to write ordinances that would permit *that*."

Last Harvest

Thanks to the failures of zoning, we hide our communities behind landscaped berms.

Tim Cassidy works for Tom Comitta as an architectural designer and landscape architect. He also happens to serve on the Londonderry planning commission. Since I'm interested in learning more about the township and the Wrigley tract, I arrange to visit him one Sunday morning. A tall, intense man of about forty with a ponytail, he's in the middle of renovating his home, a rambling yellow farmhouse. His two young daughters play among the sawhorses and the construction debris. His wife, Carolyn, holds little Rebecca, born last year. He is the eighth generation to live in the area. "My great-grandfather is buried in a graveyard down the road," he tells me. "When my parents were children, they lived across the street from one another. I was born less than two miles from this house. How's that for provincial?"

Outside, he shows me the line of white pines — fifty of them — as well as spruce and viburnum that he has planted to hide the new housing subdivision that is being built in the field behind his home. The young trees barely screen the unfinished houses and the raw, exposed earth. "I now have a new crop of vinyl where the corn used to grow," he says. Cassidy doesn't like developers.

"Ten years ago Londonderry was not a real estate market," he tells me. He says that when he was growing up, he felt removed from the urbanization of Wilmington and Philadelphia. In fact, Wilmington is only a twenty-five-minute drive—his wife, a veterinarian, commutes there to work—and King of Prussia, with its vast shopping mall, surrounding office parks, and convention hotels, is only thirty miles away. As the region continues to attract tens of thousands of jobs—QVC is headquartered in West Chester, Vanguard, with eight thousand employees, is in nearby Valley Forge—home builders and developers have turned their attention to southern Chester County, including Londonderry.

It was after a developer bought the field behind his house that Cassidy started going to township meetings. Thanks to his professional background, he found himself taking an increasingly active role in discussions. With his deep family roots in the area, he was quickly accepted by the locals and was invited to serve on the planning commission, which reviews all real estate development proposals and land-use issues for the township.

Local responsibility for land control varies widely across the United States. Constitutionally, states control land use, but except for a handful that have formal statewide zoning controls or some degree of statewide control over land use, most states—including Pennsylvania—have devolved regulatory powers over land to local governments.* All American states are divided geographically and politically into counties (called boroughs in Alaska and parishes in Louisiana), a practice derived from the age-old English shire system, but the power of counties varies considerably. In the South and West, counties control land use as well as, in many cases, schools, libraries, hospitals, law enforcement, and judicial administration. In the Northeast and much of the Midwest, counties have much less power, and control over land use resides in much

*Hawaii has statewide zoning, and both Oregon and Florida exercise state control over land use. Maine, Rhode Island, New Jersey, Tennessee, Vermont, Georgia, and Washington (and to some extent Wisconsin and Minnesota) also have a degree of statewide zoning control.

smaller units, called townships. In New England, counties have no real power. Control of land is in the hands of local municipalities, which resemble townships but—confusingly—are called "towns" and can be as large as sixty square miles.

Pennsylvania, with more than 2,600 local governments, is an extreme case of so-called home rule. Whether such a system is a good or a bad thing depends on your point of view. On the one hand, home rule satisfies local property owners, since their voices are more likely to be heard on issues such as zoning and land use. On the other hand, it frustrates advocates of regional planning, who are obliged to deal with many different—usually parochial— constituencies. It also complicates the lives of real estate developers, since each township has its own priorities and ways of doing things.

Londonderry Township was founded in 1734, when a group of Scotch-Irish settlers voted to break away from Nottingham Township. Over the years, Londonderry itself was subdivided until it reached its present size of twelve square miles. Its sixteen hundred inhabitants live in scattered houses, farms, and small residential subdivisions; there are no towns or even villages. But the small population is far from homogenous. According to Cassidy, there are four distinct groups, each with its own attitude to development. The first he calls the "old-guard farmers." Although Chester County was once entirely agricultural, it is being rapidly urbanized. In 1994 *The Wall Street Journal* listed the county among "America's twenty hottest white-collar addresses," the fastest-growing, wealthiest, and most educated concentrations in the country.[1]* *Philadelphia* magazine has called the county the successor to the Main Line.[2] The part of the county closest to Philadelphia is the most suburban; the rest is more rural but changing fast. Londonderry, for example, looks like a farming area,

*All twenty "hot" areas, which include places such as Fort Bend County, outside Houston, and Douglas County, south of Denver, are rural counties.

but farmers make up only 10 percent of the population. Cassidy describes them as free-marketers when it comes to property rights. "They would prefer no development, but if it is to happen, they want the option of selling their land." This transaction is sometimes referred to as "the last harvest."

Another 10 percent are wealthy landowners who live on large estates and are devoted to what the University of Pennsylvania anthropologist Dan Rose has called "the culture of the horse."[3] This way of life includes raising Thoroughbreds, attending polo matches, and taking part in horse shows. The cornerstone of the horse culture in Chester County is the foxhunt (of which no fewer than eleven survive). A successful hunt requires a special landscape: not simply flat fields but a combination of rolling meadows, farmland, woods, and copses, divided by jumpable rail fences. Since the riders and hounds go for miles—a foxhunt can last six to eight hours—foxhunting relies on cooperative landowners who will allow the hunt to cross their fields. In other words, not only is foxhunting an intensely social activity, it requires a high degree of cooperation among its adherents.

The modern horse culture came to Chester County in 1912, when W. Plunkett Stewart, a wealthy Baltimore securities trader, bought several thousand acres in the area, some of which he resold to wealthy friends who shared his enthusiasm for fox-hunting.[4] In 1945, when one of his neighbors, Lamont du Pont, head of the vast chemical company, put his five-thousand-acre holding on the market, Stewart arranged for his friend Robert J. Kleberg, owner of the vast King Ranch of Texas, to buy it. Later enlarged to nine thousand acres, the land was used as a fattening range for cattle shipped from the West.[5] "When I was a school-boy," says Cassidy, "the fathers of some of my classmates were cowboys—Stetsons, chaps, and all."

By the nineteen eighties, it no longer paid to ship cattle from Texas to Chester County, and the King Ranch put its grazing land up for sale. The rumored buyers included the Disney Company, which was said to be looking for a theme park site, and James W. Rouse, the developer of Columbia, Maryland.[6] To preserve their countrified way of life, the surrounding landowners turned to the local Brandywine Conservancy, which, aided by unnamed investors, bought the ranch. The conservancy added conservation easements, which forbade further subdivision, and sold the land in parcels of hundreds of acres. Thus began a scheme that eventually put 37,000 acres of land, more than two hundred square miles, under permanent protection—protection from development, that is.

Cassidy offers to show me this acreage. We drive through a spectacularly beautiful landscape of slightly rolling hills and sloping meadows interspersed with small stands of trees. I haven't seen countryside like this since I visited England; the swelling pastures and tree clumps could have been laid out by Capability Brown. Some of the estates go on for miles. As in Nantucket, wealth and preservation go hand in hand. Cassidy sardonically refers to the large landowners as the equestrian elite and characterizes them as born-again environmentalists. "They are essentially antidevelopment, or will accept as little as possible," he says. He appears torn on the subject. On the one hand, it is thanks to the equestrian set that so much of the landscape of his childhood is intact. On the other hand, he can't help being cynical about the preservation of so-called agricultural land as a playground for the rich. "But I'm really just jealous," he jokes. "I wish that I had a trust fund and could spend my time riding horses."

Eighty percent of the population of Londonderry are neither farmers nor foxhunters. Some are what Cassidy calls "old suburbanites," working people who settled here in the nineteen sixties and seventies, when land was inexpensive. Many are now retired. Their modest ranch houses and split-levels are in stark contrast to the larger and more expensive houses of the "new suburbanites," who have been moving here during the last decade. Cassidy says that there is not much mixing between the two groups. The old suburbanites attend public meetings, get involved in community activities, and volunteer for committees; the new suburbanites keep pretty much to themselves. But if the old and new suburbanites don't have much in common, they do agree about one thing: they don't want any more development.

A township's most powerful legal tool in regulating land use is zoning.[7] Zoning ordinances consist of two parts: first, regulations describing different land uses, the minimum size of lots, how much of the lot can be covered by buildings, and so on; second, a map showing how these uses are distributed in the township. Londonderry, for example, is cut up into roughly a dozen so-called zoning districts, defined according to agricultural, residential, and commercial uses, and various combinations of the three.

The first American city to adopt zoning was Los Angeles.[8] In 1908, after experiencing runaway growth thanks to several real estate booms, and citing the fire hazards of unregulated use of land, the city created zoning ordinances that distinguished between residential and industrial uses. Such legislation seemed to fly in the face of individual freedoms, but the U.S. Supreme Court upheld the constitutionality of these regulations.[9] In 1916 New York City introduced a comprehensive zoning ordinance. After that, with federal encouragement, the concept spread quickly. Within a decade, almost all states adopted laws that enabled local governments—cities, counties, boroughs, and townships—to enact zoning ordinances.

According to a textbook on zoning, states empower local governments to zone "for the purpose of promoting health, safety, morals, or the general welfare of the community."[10] That sounds sensible—having a glue factory in the middle of a residential neighborhood is not a good idea. However, the popularity of zoning is also explained by its secondary effects. Since zoning governs the minimum size of lots, it represents an effective way to control the size of the population—larger lots mean fewer residents. Since large lots generally cost more than small lots, mandating lot size is also an indirect but effective way of defining the type of person who can—and cannot—afford to live in a particular community. This is referred to as exclusionary zoning.* Zoning also has a physical dimension. Requiring very large lots, for example, means that houses will be far apart, reducing their visual impact on the landscape.

If most people in Londonderry don't want development, why doesn't the township adopt stricter zoning? I ask Tom Comitta. "I'm often hired to draft zoning ordinances that will prevent development," he says. "This can be done to some extent by enacting environmental restrictions on wetlands and slopes. It used to be common to limit development to as few people as possible by mandating large lots. But it's no longer that simple." Pennsylvania is one of the many states whose judiciary has imposed limits on home rule. In the early nineteen sixties, a developer brought a suit against a Chester County township that had raised the minimum allowable lot size on his land from one acre to four acres. The Pennsylvania Supreme Court ruled in his favor, on the grounds that a township could not insulate itself from urbanization by mandating larger lots. "A zoning ordinance whose primary purpose is to prevent the entrance of newcomers in order to avoid future burdens, economic or otherwise, upon the administration of public services and facilities can not be held valid," observed one of the Justices.[11]

Comitta refers me to other important zoning cases. In a 1970 ruling concerning an urbanizing township, the Pennsylvania

---

*From the outset, zoning had a socially restrictive aspect. Los Angeles ordinances prohibited hand laundries, for example, which just happened to be largely owned by Chinese; elsewhere, zoning was explicitly aimed at excluding African-Americans. Racial zoning was struck down by the Supreme Court in 1917, although it took more than thirty years for the Court to rule that it was unconstitutional.

Supreme Court called residential lots of two and three acres "a great deal larger than what should be considered as a necessary size for the building of a house, and therefore not the proper subjects of public regulation. As a matter of fact, a house can fit quite comfortably on a one-acre lot without being the least cramped."[12] The unstated implication is that, if a township wants to avoid legal challenges, it should make sure it permits at least some lots that are one acre or smaller.

Reading the court rulings proves more interesting than I expected. Unlike most writing about town planning and architecture, which is concerned with how things *should* be, the law baldly confronts the world as it is. In one case, the Pennsylvania court found that a township could not refuse out of hand the request of a developer for a variance to build apartment buildings even though the land was zoned solely for individual houses. "Perhaps in an ideal world, planning and zoning would be done on a regional basis, so that a given community would have apartments, while an adjoining community would not," stated the majority opinion. "But as long as we allow zoning to be done community by community, it is intolerable to allow one municipality (or many municipalities) to close its doors at the expense of surrounding communities and the central city."[13] In other words, communities cannot insulate themselves from the world around them.

The result of such legislation is that townships in southern Chester County are forbidden from stopping development outright. What they can do is to draw out the permitting process, throw up environmental roadblocks, and grudgingly—and slowly—comply with the letter of the law, as Londonderry has been doing with the Wrigley tract. In any confrontation with a developer, a township has four important advantages. First, since the courts review zoning on a case-by-case basis, the burden of proof is on the developer. Second, lawsuits take time. This is not a problem for the township, but since the developer has money tied up in the project, time is generally his enemy. Third, most small developers are loath to sue, since legal action is expensive and tends to sour relations with the community. Last, since the developer usually doesn't buy the land until permitting is complete, his commitment to a project is generally tentative, and in the face of protracted opposition, he will usually forfeit the option fee and try his luck elsewhere, which is exactly what the township is hoping for.

Despite Dick Dilsheimer's problems on the Wrigley tract, Londonderry Township is not known as exceptionally obstructive. The supervisors and the planning commission do not create artificial delays or throw up roadblocks in the face of development. They have worked diligently to create a balanced zoning plan. The northern half of the township, which contains most of the large estates, requires residential lots to be at least four acres. The southern half generally allows two-acre lots and contains several districts where lots can be as small as half an acre and where attached houses, like those being proposed for Honeycroft, are permitted.* It looks logical on paper, but the results have not been satisfactory. Subdivisions are scattered across the landscape without rhyme or reason. It has gotten so bad that each new development proposal is greeted with dread by the planning commission. "We discuss environmental issues, runoff, the percentage of impervious cover, and other technical questions, but basically, nobody likes the way these new things look," Tim Cassidy tells me after we finish our drive. "Our half-baked solution is to insist that developers build landscaped berms around their projects, so we won't have to look at them."

The situation in Londonderry is hardly unusual. Many rural communities across the United States are experiencing the same growing pains. The problem is not simply the pressure of devel-

---

*Londonderry also has zoning districts for mobile homes, shopping strips, agricultural business, and commercial-industrial uses.

opment; after all, from the beginning America has been characterized by expansion and population growth. The problem, rather, is the lack of effective ways to manage growth. Single-use zoning has proved to be notably unsuccessful in organizing the environment, since it does not address the three-dimensional nature of our physical surroundings and instead reduces everything to a crude technical measure. No wonder the popular idea of planning is simplistic: high density bad, low density good. Except that scattering houses over the landscape is *not* good. That is what Londonderry Township has discovered, and that is why they want to try something different.

I ask Cassidy about the Wrigley tract. He explains that the township requires developers to submit fully engineered plans, together with a fee to cover the cost of review. Once an application is filed, the township is obligated to render a decision within thirty days. To get a reading of the commission's preferences—and save money—developers often ask their planners to prepare preliminary drawings and present these at public meetings. In such cases, the township is under no obligation to respond quickly. "If a developer is willing to submit more alternatives, that's fine with us," says Cassidy. "That's what Dilsheimer has been doing. Because we've been focused on Honeycroft during the last few months, we haven't paid much attention to his project, except that we aren't keen on what he's proposing.

"The Wrigley tract project dragged on for several months," Cassidy explains. "We weren't getting anywhere. Finally, at our last meeting, Dilsheimer told us he was withdrawing from the project and introduced Joe Duckworth." Cassidy isn't fully converted to Tom Comitta's ideas about neotraditional neighborhoods. He's still smarting from the subdivision going up behind his house, so he's wary of new projects, but he says that he is willing to give a different approach a try.

Life, Liberty, and
the Pursuit of Real Estate

*How America's first mega-developer hobnobbed with
Founding Fathers, amassed six million acres, and landed in
debtors' prison.*

Tim Cassidy talks about real estate development in his area as
if it were something new. But it was a real estate transaction
that was responsible for the British colony of Pennsylvania, of
which Chester was one of the three original counties. Charles II
granted the 30 million acres to William Penn to settle a £16,000
royal debt to Penn's father. A devout Quaker convert, and something of a visionary, Penn maintained that the new colony was a
"holy experiment," but his description of Philadelphia is distinctly secular: "The Improvement of the place is best measured
by the advance of Value upon every man's Lot," he wrote. "I will
venture to say that the worst Lot in the Town, without any
Improvement upon it, is worth four times more than it was when
it was lay'd out, and the best forty." He fretted that "it seems
unequal that the Absent should be thus benefited by the Improvements of those that are upon the place."[1] Penn was being disingen-
uous for, as Proprietor, he was by far the largest absentee owner
in the colony. "A Map of ye Improved Part of Pennsylvania"

shows that, in addition to four choice tracts he reserved for himself, there were several others in the names of his wife and four children.[2] These properties, covering tens of thousands of acres, were grandly referred to as "manors," but they were really land banks, to be subdivided and sold as the colony grew.

In 1682, ever on the lookout for promising real estate, Penn deeded fifty thousand acres in southern Chester County to his relative Sir John Fagg, a Sussex baronet, to hold in trust for Penn's children.[3] Here is a contemporary description of one tract slated for his only daughter, Leticia:

> Beginning at a white oak at a corner of Wm Penn's manor; thence East 606 perches, part by vacant land and part by land of David Kennedy, to a post; thence South by vacant land 1400 perches to a marked tree, and West 606 perches to a white oak; thence North by vacant lands, Israel Robinson's land and Wm. Penn's Manor, 1400 perches to the beginning.[4]

A perch, or rod, is equal to five and a half yards, so the property, which came to be known as Fagg's Manor, measured about four and a half by two miles. The six thousand acres would have encompassed about a quarter of present-day Londonderry Township, including Tim Cassidy's house as well as the Wrigley tract.

Leticia, a vivacious girl, accompanied her father on his first visit to the colony but spent only two years in Philadelphia before returning to England. Over time she sold much of her American landholdings but never managed to find a buyer for Fagg's Manor. As happened often in the colony, squatters settled on the land, ignoring the absentee owner. "There is not enough left for one plantation," Leticia complained to her brother William, "wch I think very strang there is no Law to hinder such things."[5] Fagg's Manor survives today as the name of a rather forlorn crossroads, the site of a Presbyterian church and the cemetery where Cassidy's great-grandfather is buried.*

The story of Fagg's Manor is a reminder that the American wilderness was opened not only by pioneers and settlers but also by real estate developers, or land jobbers, as they were then called. The surveyor was a key member of the jobber's team, and in many cases a shareholder in the venture. Thomas Jefferson's father was a surveyor and a participant in many so-called land companies. So was the young George Washington, who early in his career was part of a company that controlled half a million acres west of the Alleghenies. In the seventeen seventies, Washington embarked on a real estate project of his own in the Ohio River country. "It may be easy for you to discover that my plan is to secure a good deal of land," he wrote to a friend who was also his land agent. "You will consequently come in for a very handsome quantity; and as you will obtain it without cost or expenses, I hope you will be encouraged to begin your search in time."[6] Washington was not successful. "I have found distant property in land more pregnant of perplexities than profit," he reflected later in life. "I have therefore resolved to sell all I hold on the western waters, if I can obtain the prices which I conceive their quality, their situation, and other advantages would authorize me to expect."[7] There were few takers, however, and when he died, the first president still owned more than 35,000 acres on the Ohio.

The British crown did not view frontier land development favorably, fearing it would aggravate conflict with Indian tribes. In 1763 a royal edict forbade the colonial governors from issuing land grants in any areas west of the headwaters of the rivers that flowed into the Atlantic and prohibited individuals from buying

---

*Following the Declaration of Independence, the Commonwealth of Pennsylvania seized the 22 million acres owned by the Penn heirs, paying £130,000 compensation, though only for the manors.

WITOLD RYBCZYNSKI

land directly from the Indians. The edict enraged people such as Washington (whose clandestine land acquisitions flew in the face of the edict), as well as Patrick Henry, George Mason, Samuel Adams, William Byrd, and Silas Deane, who were all involved in frontier land deals. It's little wonder that, when Thomas Jefferson drafted the Declaration of Independence, he included an explicit reference to land development in his list of grievances against George III.

After the conclusion of the Revolutionary War, the United States was poor in everything except land, so real estate development became the chief form of entrepreneurship. Anticipating the arrival of immigrants from Europe, local businessmen amassed vast holdings. The leading land jobber of the period was John Nicholson, now largely forgotten but a figure of consequence in the new republic. Little is known of his youth, except that he was born in Wales and came to America as a boy. When war with England was declared, the eighteen-year-old Nicholson joined the army, where he rose to the rank of sergeant. Then, in 1778, he was seconded to the Continental Congress in Philadelphia and made a clerk in the Treasury. He must have shined, for after only two years he left federal employ to become auditor of accounts for the Commonwealth of Pennsylvania. In 1782, still only twenty-five, he was appointed comptroller general, the chief financial officer of one of the largest states in the union.[8]

Philadelphia was then the federal capital, and Nicholson knew many of the political leaders of the time: Alexander Hamilton, Aaron Burr, James Monroe. Nicholson was a man of parts, a major financier, and a founder of the Bank of Pennsylvania. Charles Willson Peale's portrait, painted about 1790, shows a calm and confident figure, slightly pudgy, soberly dressed, and surrounded by account books. His placid poker face betrays not a hint of the political turmoil that is swirling around him. Nicholson had associated himself with the Anti-Federalist cause. Unluckily for him, Pennsylvania politics were dominated by

Federalists, whose stated aim was to unseat the powerful comptroller general. After several years of infighting, he was impeached. Following a twenty-three-day trial, the Pennsylvania Senate cleared him of all charges, but the House continued to insist on his removal. The governor, a Federalist but also a friend and business associate, dithered. Finally, after twelve years of service, Nicholson resigned.

He was now free to devote himself fully to his considerable real estate interests. Nicholson was clearly a bit of a rogue. His official duties had involved selling confiscated real estate and distributing land to war veterans, and he took advantage of inside information on upcoming land sales and on the availability of choice tracts to become a major land jobber. (Many veterans preferred cash to land, and Nicholson obliged.) Although he sometimes functioned as a speculator, buying and holding cheap land in the hope that he could later sell it at a profit, Nicholson was also an entrepreneur. Like a modern developer, he sought to increase the value of his properties by building improvements, in his case, turnpikes and canals. He was the early equivalent of a venture capitalist, developing an industrial town on the Schuylkill and investing in mills, copper mines, and ironworks. To promote settlement, he imported craftsmen from Europe. Since capital was scarce, like most American land developers, he looked for European investors. Through his agents James Monroe, then minister to France, and the statesman Gouverneur Morris, Nicholson sold 200,000 acres to French families fleeing the Reign of Terror.[9] When the political situation settled down and the aristocrats returned home, Nicholson resold the land to an English group of Unitarians.

Nicholson's chief business partner was Robert Morris, a prominent Philadelphia banker who had been a member of the Continental Congress and a signer of the Declaration of Independence. Morris is sometimes called "The Financier," since during the Revolutionary War he was personally responsible for raising the funds that allowed George Washington to wage his successful campaign against the British. Morris's patriotic service had not advanced his business interests, however, and he saw land development as a way to rebuild his fortune. One of Nicholson and Morris's largest real estate deals was in the nation's new capital. The federal government, which had an ambitious master plan but no funds, intended to raise money by selling individual lots. A public auction, presided over by Washington, Jefferson, and James Madison, was a dud; after three days, only thirty-five lots were sold. There was too little demand, and prices—$300 per lot—were unrealistically high. Nicholson and Morris, in partnership with James Greenleaf, a New England businessman, stepped in and made a bid for more than seven thousand lots. They offered hard terms: $66.50 a lot, payable over seven years, without interest.[10] The strapped government accepted. What one historian has called the "greatest land-grabbing triumvirate that ever operated in America" founded the North American Land Company, whose pooled real estate holdings, extending from Pennsylvania to Georgia, amounted to more than 6 *million* acres.[11]

I can't help but admire the scale of Nicholson's ambition. I tell Joe Duckworth his story and ask what personal qualities he thinks a developer needs. "Perseverance is certainly important," Duckworth says. "You also have to be ambiguity-tolerant and risk-tolerant. You must accept that you're not really in control of anything. I'm not a gambler, I don't go to Las Vegas, but I don't mind taking a risk when the odds are on my side. Being your own boss is also a part of it. The developers I know are independent and don't like the constraints of working in large organizations. And they are all extremely self-confident." That is something that Duckworth and Nicholson have in common—not only personal ambition and the willingness to take chances but also an unalloyed optimism about the future. The desire to make money, while hardly unimportant, seems secondary.

Nicholson saw a limitless future for the North American Land

Company. He boasted to prospective shareholders that the company's holdings, which were valued at fifty cents an acre, could be expected to bring in up to one hundred pounds an acre. A sales brochure made it sound easy:

> The proprietor of back lands gives himself no other trouble about them than to pay the taxes, which are inconsiderable. As Nature left them, so they lie till circumstances give them value. The proprietor is then sought out by the settler who has chanced to pitch upon them, or who has made any improvement thereon, and receives from him a price which fully repays his original advance, with great interest.[12]

Nicholson and Morris, who eventually bought out Greenleaf, believed that they could corner the market in land. But America was simply too big. However many acres they owned, there was always more cheap land over the next hill. In addition, they were unlucky in their timing. Their grand scheme was predicated on substantial immigration, but the Napoleonic Wars, and the weak European economy, slowed the flow of settlers to a trickle. The Washington venture, too, faltered. Building a new city from scratch proved more difficult than they anticipated, and they were unable to meet even the minimal obligations that the federal government imposed.

A developer can make a lot of money, or he can go bankrupt, Joe Duckworth had told my students. By the end of the seventeen nineties, Nicholson and Morris were in serious financial trouble. The carrying charges on their lands—the taxes, surveying costs, and legal fees—turned out not to be "inconsiderable" after all. They had borrowed extensively to buy the land; soon they were paying creditors in company shares. In 1798 Robert Morris wrote to his partner: "My money is gone, my furniture is to be sold, I am to go to prison, and my family to starve, good night."[13] He spent three and a half years in Philadelphia's Prune Street debtors' prison—"the hotel with the grated doors" he called it—where he received many distinguished visitors, including George Washington. His friend Gouverneur Morris (no relation) arranged an annuity for his family, which did not starve after all.

Nicholson staved off financial ruin for two more years. Ever the optimist, he told a friend in trouble, "Where you have one demand against you, I have at least 900, yet I call in the aid of reason, and do not let my mind be unfitted for business at a time when it ought to be fitted."[14] Finally, in 1799, owing the spectacular sum of $12 million, with liens on his personal property and with 125 lawsuits against him, Nicholson declared bankruptcy and was sent to debtors' prison. He died behind bars a year later, only forty-three.

Just before being sent to jail, Nicholson founded two more land companies. What appears to motivate developers, then and now, is the art of the deal. They see a difficult situation—empty backlands in Nicholson's case, recalcitrant municipalities in Duckworth's—and devise solutions by their own wits. "I love the idea that the opportunities are out there for everyone to see," Duckworth tells me. "For me, the deal is not simply I do this and you do that. It's a chance to be creative, to get the other party to be satisfied in a new way. I like the challenge of getting novel kinds of entitlements, or convincing a builder to try something unusual. In a good deal, everyone is better off."

Joe's Deal

From a developer's point of view, the United States is divided into two areas, not rich and poor, or rural and urban, but pro-growth and anti-growth. The South, Southwest, Texas, and Florida embrace development, while the Northeast, Northwest, and California discourage anything that threatens change.

Tim Cassidy had given me driving directions to the Wrigley tract, which is only five minutes from his house in a place called Daleville. There is no sign, and I almost miss it. Daleville turns out to be a country crossroads with a battered telephone booth and a closed-up convenience store, whose large display window is curtained by a faded American flag. Across the road is the site: a mixture of cornfield and pastureland sloping down to a small woods. It's a wonderfully bucolic scene. Chestnut Hill in 1880 must have looked something like this, but it's still a surprise—I wasn't expecting to be surrounded by open countryside.

The Wrigley tract is the smaller of two adjoining farms that once belonged to Wayne Wrigley, a local farmer who bought them in the nineteen thirties. When he retired, in 1964, he sold the larger, 175-acre parcel to his elder son, Marvin, and moved to the smaller farm. After Wrigley's death, this passed to his younger son, Wayne Jr., a radiologist. When Dr. Wrigley, who divides his time between Daleville and Tampa, retired, he put his 90-acre farm up for sale, keeping only the house and some surrounding land. That's where Dick Dilsheimer came in.

What is the value of the Wrigley tract? According to Joe Duckworth, as a cornfield it's probably worth about $300,000. Since the zoning allows residential use, a developer might pay between $600,000 and $1 million, which could be said to be the current market value. Dilsheimer has offered considerably more than that—$1.88 million. His offer is contingent on the township granting permission to build within two years; otherwise the agreement will be void and the seller will keep the $60,000 deposit. Duckworth calls the difference between the market value and $1.88 million a risk premium, that is, in return for the extra payment, Dilsheimer has a two-year option on the land.

According to Duckworth, the economics of land development are simple: "You spend money buying and subdividing land, then you make money selling lots." His back-of-the-envelope calculation for the Wrigley tract is equally straightforward. He figures that in the current market he will be able to sell house lots to builders for about $60,000 each. On the debit side, he will have to pay $1.88 million for the land, and probably something to Dilsheimer as a finder's fee. Duckworth estimates that the infrastructure will cost $30,000 per lot and that the entitlement process, with its consultant fees and permitting costs, will add another $5,000 per lot. Typical profit margins for this kind of development are between 15 and 25 percent. To make a conservative profit of, say, 17 percent, Duckworth calculates that he will have to sell at least 125 lots, rather than the 86 that Dilsheimer planned.

"We'll do more accurate estimates later," says Duckworth. "Business plans are never right on, they always tend to be low. But the bottom line is probably right." He gives me a brief explanation of land development economics. From a developer's point of view, the United States is divided into two areas, not rich and poor, or rural and urban, but pro-growth and anti-growth. The South,

Southwest, Texas, and Florida embrace development, while the Northeast, Northwest, and California discourage anything that threatens change.[1] "Considering that Philadelphia is growing very slowly and Atlanta is booming, which area do you think has higher new-house prices?" Duckworth asks. "Conventional economic theory would say Atlanta, but prices of new houses in Philadelphia are twice as high. You can see the same difference between metro Detroit and metro Dallas. The explanation is the permitting process. In the South it takes two to three months to get a permit, while in the Northeast it takes two to three years." So why aren't you in Atlanta, instead of Philadelphia, I ask. "In Atlanta, competition is fierce, so profit margins are lower," Duckworth explains. "In the Philadelphia area, entitlements are so hard to get that, when you do get them, you know that you won't have a lot of competition. Supply is constrained, and builders are desperate to find permitted land. So, the price of lots is almost guaranteed to keep going up."

That's why Duckworth expects that, although his costs will eventually be higher than the first estimate, by the time the project is built, land prices also will have gone up. If they rise enough, he may even increase his profit margin, but he isn't counting on it. For the moment, assuming things go smoothly, the Wrigley tract, while not promising a windfall, looks like a reasonable bet. There is a major obstacle, however. The current zoning allows a maximum of ninety-seven single-family houses on half-acre lots. Duckworth needs to sell more lots, and he wants to make them smaller, to create the feeling of a village. Before finalizing his deal with Dick Dilsheimer, with whom he has a handshake agreement, Duckworth must be sure that the township will approve such a major zoning change. To do that he needs a plan.

The person responsible for planning Duckworth's new community—laying out the streets, lots, and open spaces—is Bob Heuser.

The city planner is a large, easygoing man in his early fifties, with swept-back, prematurely graying hair and a bushy mustache. He works in an office in his home in suburban Philadelphia. The large room overlooks a steep wooded slope. "Everyone's on lunch break," he says, pointing to the empty work area. He's joking—Heuser works alone. The airy room contains a small conference table, a secretarial desk, large flat drawers for filing drawings, and an old-fashioned drafting table. Heuser does not design on a computer. "I enjoy drawing," he says. "I work things out in my head, and my hand just follows along."

Heuser's clients are mainly suburban developers. One of his regulars is Dick Dilsheimer, who hired him to plan the Wrigley tract. "The first time I met Dick was on a project where he had been unsuccessfully trying to get township permission to build a hundred town houses on a piece of land for *twenty-five years,*" Heuser says. "He's a patient man. But after only a few months working on the Londonderry project, I could see that he was getting nervous. Then we heard that the township might be open to a village-style plan." Heuser had designed many projects for Dilsheimer over the years and knew that his cost-conscious client was not interested in radical planning innovations. "His philosophy is 'People want Buicks, we make Buicks.'" Heuser designed what he calls a faux-village plan: the same number of houses but on slightly smaller lots. The reaction from the Londonderry planning commission was lukewarm. "So I came up with a suggestion," he says. "I told Dick that he should contact Joe Duckworth, who I knew was interested in unconventional projects. A short time later, Dick called me and said, 'Bob, you may have a new client.'"

Heuser has known and worked for Duckworth for more than thirty years. "Joe is an interesting guy," he says. "He really believes in the neotraditional concept, people-friendly environments, walkable communities, and the like. But he's also a developer. He knows that to stay alive he's got to make money." Heuser has thought a lot about traditional neighborhood development. "I would say that about fifteen percent of my projects now fall into this category." He has not been to Seaside, but he has visited similar projects. Recently, he has been working on several neotraditional plans with Tom Comitta. While he appreciates Comitta's enthusiasm, Heuser isn't ready to buy into the neotraditional neighborhood philosophy. "Right now it's the flavor of the month," he says. "I come at the subject from a different perspective. I'm an advocate of livable communities, but they don't necessarily have to be in the neotraditional mold, as long as they're pedestrian-friendly and well-planned."

Heuser shows me his new plan for the Wrigley tract. He has designed several neotraditional projects for the Arcadia Land Company, so he knows what Duckworth wants. He has laid out 135 small lots, each only five thousand square feet in area, fifty by one hundred feet, less than one-eighth of an acre. Instead of the seventeen acres of open space in Dilsheimer's project, there are now more than fifty. The houses are grouped together, with rear garages and lanes. Some of the lots back onto the old Wrigley farm, now owned by Charlotte Wrigley, Marvin Wrigley's widow. After her husband's death in 1977, she and her two sons granted an agricultural easement on their farm, which ensures that the 175 acres will remain undeveloped in perpetuity. Heuser figures that these perimeter lots will sell at a hefty premium. "Joe calls it the rural equivalent of an ocean view," he says.

When Duckworth first presents Heuser's plan to Londonderry Township, the planning commission likes the additional open space but is concerned that farm operations and residences will make poor neighbors. People like the idea of living in the country, but they complain about farmers spreading fertilizer or working late at night. The commission asks to see a plan with the houses relocated on the eastern side of the site. This doesn't appeal to Duckworth. Most prospective buyers will approach the new development from the east, so having an attractive open space on that side makes marketing sense. He suspects that the change in location has something to do with Charlotte Wrigley, who happens to be the township secretary, wanting the development as far away from her farm as possible. Nevertheless, he instructs Heuser to go along with the suggestion. He also agrees to a request to reduce the number of houses to 125—which is the number he was aiming for anyway. The important thing is that the planning commission seems to have accepted the general idea that there are going to be more houses on smaller lots.

Heuser makes a second plan. Nobody likes it. Not only is it less attractive—actually, the way he has drawn it also *looks* less attractive—than the first version but the houses are now close to Route 796, a busy country road. The commission asks Duckworth to go back to the first plan but to create a buffer of open space between the houses and the Wrigley farm. He happily agrees. A final version of the plan takes shape on Heuser's drawing board.

This back-and-forth takes five months, and Duckworth is still far from the point where he is ready formally to submit a preliminary plan. But he now feels confident enough to finalize his agreement with Dick Dilsheimer. He pays him $110,000, which represents the money Dilsheimer has already spent on the project, and promises to pay him an additional $125,000, that is, $1,000 per lot, at the time of the lot's sale. Dilsheimer, who has devoted almost a year to the project, is happy to be out of it. "Joe has his own vision," he says. "We don't share it. Our buyers like their forty-acre spreads."

Dilsheimer is exaggerating, but he's right that people like space. A 2004 study, carried out in Columbus, Ohio, asked twelve hundred homeowners a series of questions about hypothetical housing preferences such as neighborhood layout (conventional versus neotraditional), density, and the influence of parks and open space. The researchers analyzed people's preferences in the context of randomized combinations of potential neighborhood characteristics. "Despite the popularity of [neotraditional] projects among planners and urban designers, our survey results indicate less interest on the part of the home-owning public," the authors concluded. "In general and all else being equal, people prefer low density."[2]

This sounds like bad news for Duckworth's project. However, the Ohio study also concluded that there was a small market for village-style developments: "Although our results indicate that on average the lower-density choice was preferred, we found that the mean respondent would choose the higher-density neo-traditional-style neighborhood under a range of plausible conditions."[3] These conditions include the presence of a park, shorter commuting time, and the adjacency of land in permanent agriculture. The Wrigley tract has most of these characteristics, so while Duckworth knows that popular demand for smaller lots and higher density is probably limited, he believes it will be large enough to make his project successful. He negotiates a revised contract with Dr. Wrigley, the landowner. Since the project is, in effect, starting afresh, Duckworth needs more time, and he asks for an extension of the option to buy, which costs him another $60,000 advance.

Dave Della Porta is an energetic man in his late thirties with dark, curly hair. He was one of the first people Joe Duckworth hired at Realen. At that time Della Porta was a young lawyer handling real estate in one of Philadelphia's big downtown law firms who wanted to get into the development business. "Joe taught me the ropes and put me in charge of buying land," he says. Della Porta quit Realen, shortly after Duckworth left, to start his own development company, Cornerstone Communities. Cornerstone's business is mostly building low-rise apartments and condominiums in the suburbs, but Della Porta also builds a small number of single-family houses and develops suburban subdivisions. Cornerstone Communities has six employees and operates only in the Philadelphia area. "I want to be able to drive to any project from my office in no more than an hour," he says.

Della Porta has collaborated with Duckworth from time to time. He has been the developer on one of Arcadia's projects and a builder in another, and Duckworth has invested in several Cornerstone developments. Della Porta will be a partner in the Wrigley tract deal. Arcadia will be in charge of buying the land and overseeing the permitting process with the township, while Cornerstone will be what Della Porta calls the "managing partner," responsible for building the roads, sewers, and other site improvements, and marketing the lots to builders. The equity for the project, about $2.5 million, will come 90 percent from Arcadia and 10 percent from Cornerstone. More than half of the Arcadia share will come from Duckworth himself, making him the majority investor.

The so-called soft cost of a development project is the cost of the various planning, engineering, and infrastructure consultants. In the case of the Wrigley tract, this will be about half a million dollars. "You never borrow up-front costs," Duckworth tells me. "That's the price of getting into the game." The main expense is the hard cost—buying the land and building the infrastructure—which will add another $7 million. Five million dollars of this will be debt in the form of a construction loan from a local commercial bank. The loan, arranged by Cornerstone with Arcadia's equity as collateral, will be repaid as lots are sold. That is the real risk. "Two or three years from now, when we start to sell our lots, there could be a downturn in the economy," Duckworth says. "If house sales slow down, builders will not buy lots, and we'll be left holding the bag for several million dollars, depending on exactly when, in the five-year process, the downturn occurs." Even without a downturn, Duckworth may not get the prices he hopes for. Londonderry is a little outside the main development areas of Chester County, and the market for first-time buyers is very price-sensitive, so the success of the project will depend on keeping costs low, and on providing an attractive development that home buyers will see as a good value.

One of the reasons Duckworth has taken a partner is that he's busy. He has a conventional development of high-end houses nearing completion in an established suburb of Philadelphia and is shepherding another neotraditional project—Woodmont—through the delicate final stages of the permitting process. At the same time, he is exploring several other possible projects, including a village cluster in nearby Sadsbury Township, as well as a low-rise condominium in downtown Wayne, a Main Line community. That is typical; Duckworth is always looking for new prospects. This is not only because the permitting process for any particular project takes so long; like all developers, he knows that many of his deals will fall by the wayside.

Duckworth has recently suffered two setbacks. The first was a development for a beautiful 483-acre tract called Betty's Neck, overlooking Assawompset Pond in southeastern Massachusetts. The land belonged to a family of cranberry growers, driven to sell by collapsing cranberry prices. The town of Lakeville wanted to buy Betty's Neck for a park but was unable to raise the funds. The family's planner contacted Duckworth to see if Arcadia could develop the site, leaving the land around the pond accessible to the public. Duckworth made a proposal for developing five hundred lots in several village layouts on half of the developable land, conserving most of the site in its natural state. He was enthusiastic about the project. Not only was this an exceptionally attractive location but it promised to be very profitable. "It's just within commuting distance of Boston. Over the life of the project, house prices will likely rise as high as six hundred and fifty thousand dollars, so you don't have to be a Wharton grad to see that it's off the charts." The planner's strategy was to present a conventional subdivision plan to the town and then unveil Duckworth's village option. But events took a different turn. A land trust, which had been encouraging the town to buy the land, stepped in. Thanks to a last-minute grant from the state, the town and the trust made an offer that the family accepted. Duckworth suspects his proposal may have been used to leverage more money from the state, but he's philosophical about it. "That's the way the game is played. The trust option is easy money, without any of the headaches of development," he says.

Betty's Neck didn't get as far as the planning stage, so Duckworth didn't lose any money. That wasn't the case with another project. Fairstead was a large development planned on three hundred acres in Lancaster County. The land belonged to Armstrong Industries, which has its world headquarters in Lancaster. The planned development was to include a village center as well as a full range of housing types, fifteen hundred dwellings in all. Fairstead had all the makings of a flagship project for Arcadia, including a sympathetic—and patient—corporate landowner who was interested in the long-term future of the site. The township was enthusiastic. After three years of planning, just as approval was to be granted, Armstrong, weighed down by asbestos-related lawsuits, suddenly declared bankruptcy. Under Chapter 11 proceedings, the contract with Arcadia was canceled, and Duckworth and his two developer partners were out almost a million dollars.

Duckworth has a personal reason for asking Dave Della Porta to become involved in the Wrigley tract. Duckworth's eldest son, Jason, has just joined Arcadia, and Joe wants Della Porta to work with him and train him, just as he was trained at Realen. Jason Duckworth, thirty years old, tall and slim, doesn't resemble his father. Nor is he an MBA. He graduated in urban studies from Princeton, where he was Phi Beta Kappa and received a scholarship to Oxford. There he learned Mandarin in the summer and got a master's degree in geography. While still in England, he was recruited by McKinsey, an international business consultant, and moved to New York City. "When I graduated from Oxford, Arcadia didn't exist yet. In any case, I didn't give much thought to working in real estate," he says. "I had an exciting, well-paid job in New York, working for a high-profile company." At Oxford he had met Angela Lee, another McKinsey recruit, and they married. In 1998 he moved to a venture capital firm in San Francisco. Angela, who had left McKinsey to teach math in a New York charter school, got a teaching job in a public school. "It was the beginning of the dot-com boom, and San Francisco was the entrepreneurial capital of the world. I was the first of my group at McKinsey to go. A year later, twenty percent of my class was there," Jason says. Through his father, Jason met Robert Davis, the developer of Seaside, and his wife, who had just moved to San Francisco. The couples became friends. "I had been an urban studies major," Jason says. "So I loved talking to Robert about town planning and architecture."

The dot-com boom peaked in 2000. Jason could see that his firm's investments were not performing as expected. "I liked the work, but I was concerned that, with my liberal arts background, I wasn't likely to be one of the stars in this more difficult environment. And Angie was expecting our first child, so I was thinking about the future," he explains. Jason knew that Arcadia was getting lots of work and thought that, with his background in urban studies and his experience in business, he could contribute to the company. "But it wasn't an easy decision. I had so much invested in the other career path," he says.

Finally, he left the venture capital firm and joined Arcadia's San Francisco office. I ask Jason if he had been influenced by working with Robert Davis. "Oh my gosh yes," he says. "Robert's story is inspirational." Jason and Davis spent a month at Seaside, where Jason worked on a plan for the sale of the final properties. He loved Seaside. "I'm a wannabe architect," he says. "I took every architecture course I could at Princeton. To be able to do something like Seaside is great. From Robert I learned that, quite apart from the money, real estate development can be really satisfying." Jason spent the next eight months in San Francisco, helping Davis on several Arcadia projects. Finally, he decided to go back to Philadelphia. His father had more work than he could handle and was thinking of recruiting an MBA from the Wharton School. "Mainly for personal reasons—both our families are in the area—it made sense to go back," says Jason.

"It was a big move for Jason to join Arcadia," says his father. "To his peers, developing residential real estate is one step above running a garage." Duckworth is obviously pleased that his son has joined him. "When I was at Toll Brothers, I was always against nepotism—I was afraid I would have to work under some incompetent person," he says. "But now that I have my own company, I've brought Jason in, and I'm trying to get my other son, the seminarian, involved."

On the Bus

Narrow streets in neotraditional neighborhoods make for slower traffic—and lemonade stands.

Before the Arcadia Land Company can present a plan for approval, Londonderry must amend its zoning. A conventional zoning ordinance is a straightforward document. The ordinance that currently governs the Wrigley tract, for example, consists of only three pages that describe the permitted uses, minimum lot sizes, and basic building characteristics, such as maximum footprint and height. A neotraditional ordinance is slightly more complicated, since it defines not only lot sizes, setbacks, street and lane widths, and the required area of open space but also various architectural features of the houses.

The new ordinance is needed to allow Arcadia to build 125 houses instead of 86. Without this increase, the development will not be economically feasible. Work on drafting the new ordinance begins at the same time that the planning commission reviews the first sketch plan. Tom Comitta has written a similar ordinance for a nearby town, and Jason Duckworth, who is now working on the project, volunteers to rough something out using Comitta's ordinance as a model. "It might sound odd to have the developer involved in this way," says Comitta, "but Jason is a bright, energetic, resourceful person, and since the township is interested in getting a successful project, we appreciate his initiative." A working group is formed that includes Comitta; Jason; Bob Harsch, the township engineer; John Halsted, the township solicitor; and David Sweet, who advises Londonderry on zoning.

Over the next three months this group meets half a dozen times, usually in Comitta's office, drafting the language of the ordinance. One point of contention between the developers and the township is the precise definition of open space. Jason feels that anything not a private lot should be considered open space, while Comitta maintains that open space has to be accessible and usable. Another issue is the width of streets. Londonderry ordinances typically call for a fifty-foot right-of-way, which means that the actual pavement is thirty-four feet wide. Arcadia wants the pavement on minor streets to be twenty-four feet, with parking on both sides as well as two narrow lanes of traffic. Narrow streets are a crucial ingredient of neotraditional neighborhood design since they bring buildings closer together, creating a more intimate and defined street. Narrow streets also slow down traffic, which makes it more pleasant—and less dangerous—for pedestrians (a person dies if hit by a car going more than forty miles per hour). According to traffic studies, a twenty-four-foot-wide street with cars parked on two sides slows traffic down to fifteen miles per hour, since cars are obliged to stop to let each other pass, something that traffic engineers call yield movement.[1] Of course, narrow streets will also mean lower construction costs for the developers. But the township is worried about traffic accidents and legal liability. The two sides finally compromise on a twenty-four-foot-wide street but with parking on only one side, which raises the speed of traffic slightly, to twenty miles per hour, but produces the street dimension that Arcadia wants.

Finally, in mid-July, the ordinance is ready for review by the township planning commission. The seven commissioners, who meet publicly once a month, represent the three established groups of the township: estate owners, farmers, and longtime residents.

"The Arcadia proposal is unprecedented," says Tim Cassidy. "Our commission is not generally well-disposed to residential development, and reducing the number of units is considered a great victory. And here is a developer asking for a fifty percent *increase* in the number of houses." Despite the earlier favorable reaction to neotraditional development, and their stated desire to try something new, it turns out that the planning commissioners are far from unanimous. One, a farmer, is dead set against anything that increases density. Another, who owns a large horse farm, is concerned about where the new development will get its water. What finally sways the debate is a comment by a man in the audience. "We've been doing conventional development and we hate it," he says. "Why don't we try something new, and if we don't like it we won't do it anymore." It is, as Cassidy, who votes in favor, puts it, an epiphany moment. With one commissioner absent, the vote is four-to-two in support of the neotraditional ordinance.

Although the final decision on the new ordinance will be made by the supervisors at another public meeting, the recommendation of the commission carries weight, so Jason is pleased with the result. A month later he learns that the situation has changed. It turns out that there was a procedural slipup. When the minutes are read at the planning commission's next meeting, they reveal that the four-two vote was not on the ordinance itself but only on a motion to advertise it. (The township is obliged by law to publish any new ordinance in advance of the supervisors' meeting.) A second vote is required. This time, one of the original supporters, a veterinarian, is away on a trip, while the previously absent member, now returned from vacation, votes against the motion. That makes it a three-three tie.

Jason fears that a divided planning commission will make it harder for the supervisors to support the project. To make matters worse, there is disturbing news from the county. Pennsylvania counties have no authority to issue permits or approvals—theirs is strictly an advisory role—but state law requires that they review any zoning changes. The Chester County planning commission has contacted Londonderry and asked for an extension to its customary thirty-day review period. The commission has some unspecified reservations about the new ordinance that they want clarified before they vote. If the county raises serious objections, there is a real possibility that a skittish board of supervisors could vote against the ordinance. "I don't think this is a cakewalk any longer," Jason e-mails Dave Della Porta.

While a vote on the Londonderry ordinance is slowly coming to a head, Arcadia is in the early stages of a project in nearby Sadsbury Township. This development requires a controversial zoning change, from industrial to residential. To build support in the community, the Duckworths have organized a bus trip to Kentlands, a large neotraditional development in Maryland. They have invited the Londonderry supervisors and members of the planning commission to come along, too.

The Sadsbury group meets behind the firehouse at 7:30 A.M. on a sunny Saturday. The developer in training Jason Duckworth is handing out paper cups of coffee from the open tailgate of his station wagon. "We have donuts, cinnamon buns, and muffins," he tells the people milling around in the parking lot. Jason is helped by Christy Flynn, a Wharton student who is an intern in the office. Joe Duckworth is there, too, wearing a baseball cap and a white polo shirt with an Arcadia logo. Everybody has a name tag, although most of the locals seem to know one another, and there is a lot of kidding around. The last time many of them took a bus was probably when they were schoolchildren, which may explain the playful atmosphere. When the bus arrives, it's not yellow but a shiny black luxury coach. Some of the Sadsbury residents exchange glances; they're impressed.

Arcadia sent out personal invitations, took out ads in the local newspaper, and made announcements at township meetings.

Eighteen people have showed up, which pleases Duckworth. "The last time I tried this sort of thing, only four people came," he tells me. Everyone climbs aboard. Duckworth briefly introduces himself, Jason, and Christy, and thanks people for coming. There is a smattering of applause, not for Duckworth but for the bus driver as he skillfully eases the huge vehicle out of the cramped lot and into the narrow village street.

Twenty minutes later, the bus stops at the Londonderry Township building, where a handful of people are waiting in the parking lot. The important thing, as far as Joe and Jason are concerned, is that one of them is Howard Benner, the chairman of the board of supervisors, and another is Richard Henryson, the chairman of the planning commission. The last person to get on is Londonderry's planning consultant, Tom Comitta, who has been to Kentlands before but wants to encourage his clients.

The bus is equipped with television screens, and Jason plays a videotape of interviews with Kentlands residents. The residents talk about why they chose to live there and how much they like it. Some say that it reminds them of a New England town, others mention the open green spaces and the lakes. Everyone has something to say about how pleasant and easy it is to walk around the community, how the kids can bicycle to school, and how you can walk to the store. The film is obviously a marketing tool, but it isn't slick and the statements sound sincere.

Comitta gets up and, from the aisle, gives a more technical explanation of the planning ideas behind projects such as Kentlands. He lists what people should look for during their tour: the small lots, the variety of types of housing, the design of the streets. He also recounts his own early experiences as a township consultant and his conversion to traditional neighborhood development. He is candid and enthusiastic. Joe Duckworth makes no more comments after his brief introduction. "I prefer to keep a low profile during these tours, and let people make up their own minds," he explains to me. "Nobody trusts the developer anyway."

Meanwhile, the big bus barrels south along I-95. It takes more than two hours to get to Kentlands, which is in Gaithersburg, a Washington, D.C., suburb. The bus pulls up in a parking lot in the town center, and the group files off. The first stop is the office of the neighborhood newspaper. Our guides for the tour, Valerie and Leslie, two attractive women in their early thirties, hand out literature about the development. Both live here. It turns out that they've actually worked for the developer, so they aren't exactly typical residents, but they come across as energetic and friendly, and they have a lot of facts at their fingertips.

Kentlands is an important milestone in the history of neotraditional design, since it was the first development that applied the concept of Seaside to a suburban community. Planned by Andrés Duany and Elizabeth Plater-Zyberk, Kentlands, with two thousand homes, is much larger than Seaside. The project began in 1987, and the first houses went on the market three years later. The early indications were positive. Kentlands, benefiting from Seaside's celebrity, received national publicity, and sales were brisk. Then, the one thing that real estate developers fear most happened: the economy faltered. This affected the development in two ways. First, home sales slowed down. Second, because of reduced consumer confidence, merchandisers were reluctant to open new stores, which meant that the town center remained unbuilt. For Joseph Alfandre, a local builder and the pioneering developer of Kentlands, who had been counting on the sale of the town center properties to finance the project, this was a heavy blow. By 1991 the economic slowdown had turned into a full-fledged recession. Underfunded and unable to weather the slump, Alfandre was forced to hand over the project to his main creditor, a local bank. Over the next several years, the bank completed construction more or less according to the original plan. By the end of the decade, as the economy recovered, sales improved and land values rose. Kentlands became a success. An economic study comparing the development with other Gaithersburg subdivisions found that, despite the smaller lots, home buyers were willing to pay a premium for the green spaces, the convenience of nearby shops, the townlike surroundings, and the comfortable sense of community.[2]

The Chester County group sets off on its tour. The physical reality more than meets the expectations raised by the promotional video. Kentlands has many of the qualities associated with garden suburbs: shady streets, a picturesque layout, houses close together. The architectural style is predominantly Federal, and the compact appearance of the brick and clapboard houses lining the streets really does recall a small New England town. So do the white picket fences. In fact, it's all so pretty that the first impression is of a place that is slightly unreal. "It's like Disneyland," someone says skeptically. But as the walk continues, impressions appear to change. It's a sunny weekend morning, and many people are outside. A woman on her porch with her baby waves hello. Residents are washing cars, cutting the grass, gardening. In front of a yellow house, in a scene worthy of Norman Rockwell, a group of small children has set up a lemonade stand — 25 CENTS A GLASS says a hand-lettered sign. If the Duckworths had wanted to stage an event for the benefit of this tour, they couldn't have done better.

I have visited Kentlands several times, but I am still struck by the sophistication of Duany and Plater-Zyberk's plan. The small open spaces break up the street grid. The houses are mixed to avoid uniformity, a few town houses nestle along a street of larger homes. The streets are subtly angled, just enough to skew the view occasionally and produce the sorts of "accidental" effects that make a place like Chestnut Hill so charming. Of course, Kentlands is a little too considered, a little too prim and, well, cute. But Chestnut Hill has had more than a hundred years to ease itself into shape, whereas Kentlands has been in existence only a decade.

Kentlands promotes the notion that it is a small town. There is a Main Street and a Market Square; the area where *The Town Paper* office is located is called Midtown. "Kentlands is actually more than a neighborhood," reads the original sales brochure.

"When completed it will be a small town with several different neighborhoods spreading over three hundred-and-fifty acres of Maryland countryside." In truth, Kentlands is not a country town but a suburban master-planned community, bounded by highways, shopping malls, and other master-planned communities. Yet children can walk to school and bicycle to the corner store. The narrow streets do create the impression of a small town, and the proximity of the houses does foster neighborliness. The so-called Midtown area has many live-work units, attached three-story buildings that allow a combination of commercial and residential uses. The mix of lawyers' and doctors' offices, restaurants, and shops creates an atmosphere that really does resemble a small-town main street.

Our group passes an elementary school and a day-care center, a hilly neighborhood of Victorian-style houses, and a clubhouse where teenagers are noisily playing in a swimming pool. It is more than an hour since the tour started. "We've been walking quite far. I hope that you're not getting tired," says our guide, concerned about the elderly members of the group. "Oh, no," jokes one of the farmers. "It's farther to go to my mailbox."

It's almost noon, and we're back at the bus. Jason announces that box lunches will be available in twenty minutes. In the meantime, he suggests, people can visit the town center. The businesses here are not much different from what one might find in any small shopping center, and in fact Market Square is owned and managed by a shopping mall operator. But the shops, restaurants, and multiscreen movie theater are in individual buildings, and instead of an indoor mall, there are sidewalks, trees, and streets. The parking lots are behind the buildings, out of sight.

It's an attractive setting, yet there are few shoppers. "The hardest thing to make work in a neotraditional development is the town center," observes Duckworth as we stand on a street corner looking at the empty sidewalks. Although Kentlands has about two thousand families, and neighboring Lakelands is approaching a thousand households, that is still not enough people to support a large number of stores. The problem is that, despite their often stated preference for walking, Americans have developed a taste for low prices and variety, and they don't mind driving great distances to get them. "Neotraditional development is definitely a good idea," says Duckworth, "but we still haven't found the right model for how to put together a successful town center. It's not Seaside, whose town center works but is really a beach resort. Nor is it Disney's Celebration, whose downtown is a tourist destination. It's certainly not this place."

It is time to leave. Back on the bus, we contentedly munch our sandwiches and sip soft drinks. On the way home, Comitta summarizes his impressions to the group and asks if there are any observations, pro or con. "Do you know how much open space there is?" asks a man from the back. Comitta isn't sure; Duckworth says he thinks that it might be 25 or 30 percent. "I like the fact that it was very clean," says one woman. "It shows pride of place," chimes in another. On the whole there is not much discussion. It has been a long day, and people are ready to go home.

Jason goes down the aisle handing out bottled water. He is pleased with the way things have gone. The Londonderry Township representatives, Benner and Henryson, seem genuinely impressed. "Nobody ever comes away from Kentlands feeling worse about our proposals," says Jason. "By the way, we think that the bus ride is an important part of the trip. It allows us to talk to people in a more intimate environment. It also gives us a chance to demonstrate to the residents that we're human, too. The public and confrontational environment of the township meetings creates a very depersonalized view of developers." His father has learned to joke about his perceived role as the "greedy developer," but for Jason, who has absorbed Robert Davis's benevolent vision, it's important for people to understand that he is interested in more than the bottom line.

8

Meetings

Why developers hate to go to public meetings.

The Londonderry board of supervisors meets on the second Tuesday of the month at seven in the evening. The township building, which is in Daleville, is a long industrial shed with a row of garage bays for trucks and road maintenance equipment. The township offices are attached to one end, a little civic afterthought.

I arrive early. The gravel parking lot is beside a large unmowed field identified by a sign as the Londonderry Township Park. Lower down the slope, on the far side of the field—I can't bring myself to call it a park—is a relatively new subdivision whose name I noticed as I drove in: Mindy Acres. Since there are no trees, no hedges, and no fences, I have a panoramic view of the entire development. Except for a strip of asphalt roadway—no sidewalks—everything is green turf, which makes it look a little like a golf course. The large houses, far apart on what look like one-acre lots, are turned this way and that, tenuously connected to the winding streets by long driveways. It's hard to characterize this artless arrangement. Mindy Acres exists in some not-quite-rural, not-quite-suburban limbo.

The parking lot fills up with pickup trucks and SUVs, and I go inside. The township hall is a large, undecorated room with a low suspended ceiling and fluorescent lights. There is an American flag in the corner. About fifty metal folding chairs are set up in rows on the carpeted floor. Most are occupied. The three supervisors, who are the township's elected government, sit at a long table at the front of the room. The chairman, Howard Benner, tall, in his seventies, has long worked for the school district; Clair Burkhart, a ruddy-faced man, owns a local excavating business; and Fred Muller, bearded, in shirtsleeves, runs a small organic farm next to Mindy Acres. Charlotte Wrigley, an elderly lady with white hair, sits at one end of the table with a notebook, while Bob Harsch, the township engineer, is at the other end. He is the only man in the room wearing a tie.

Tom Comitta bustles in with an armful of papers. Township meetings are his bread and butter. "I have sixty years experience," he jokes, "thirty years in the day and thirty years in the evening." He is followed a few minutes later by the Duckworths and Dave Della Porta. Jason is carrying a large leather portfolio, from which he removes several presentation boards that he pins up on the wall. The large plans, colored to highlight the trees and landscaping, are titled "New Daleville." "It was time to name the project," he says.[*1] "We wanted Daleville in the name, but the township didn't want us to use Daleville alone, so we brainstormed different combinations. We considered Daleville Center, Daleville Farms, and Daleville Village. My dad suggested New Daleville, and it stuck." The room is full, which for Jason is not necessarily a good sign. "But compared to some other projects Arcadia has done, this one has gone pretty smoothly," he tells me. "We think the vote is going to go our way. But I still have this feeling in my stomach that something could go wrong."

Punctually at seven, Benner calls the meeting to order. He

---

*The practice of naming residential subdivisions originated in 1811 in England, with John Nash's suburban Bristol development, Blaise Hamlet. Llewellyn Park, Riverside, and Wissahickon Heights are early American examples.

announces that the Chester County planning commission is meeting tomorrow afternoon to discuss the proposed neotraditional ordinance and has asked for a postponement of the supervisors' vote until next month. The news that there will not be a vote produces a dissatisfied grumble from the room. The law requires advance notice of zoning hearings, and people have made a special effort to come and have their opinions heard. Benner apologizes for the last-minute change and says that there is no reason why a discussion can't take place as scheduled. He asks Comitta, as the township consultant, to describe the new ordinance.

Comitta launches into a spirited explanation of the philosophy of traditional neighborhood development. He talks about the need to rethink conventional planning and criticizes dependency on cars and the large lots of conventional residential developments. He casts the proposed ordinance as a high-minded alternative to sprawling development. Warming to his subject, he points out the window to Mindy Acres. "That's the sort of thing that we're trying to avoid."

During Comitta's presentation, a bearded man and his wife who are sitting directly in front of me have been angrily whispering to each other. When Comitta finishes, the man stands up to ask a question. It appears that he and his wife are residents of Mindy Acres. "Do you mean to say that we've been doing it all wrong?" he asks incredulously. "All this time? Really?" Then he adds more belligerently, "I *want* to drive to work. That's why we moved here. To get away from the traffic and congestion." There are murmurs of agreement in the room. The audience is a mixture of people, not farmers but rather working people living in a farming area. Many appear to be "new suburbanites" who live in the immediate surroundings. Obviously, they don't see having more neighbors as a plus. Comitta's call for denser development has not gone over well.

"I moved here about ten years ago," says a mild-mannered man, "and when the house I live in was built, there was probably a meeting like this to complain about it." When the laughter dies down, he continues. "I'm not against development, but I am concerned about the extra traffic. How will this affect our standard of living? Are we going to be paying for road repairs and improvements afterwards?" There are a number of other questions along the same line. If families with children move into the area, will school taxes go up? What about the cost of policing? Couldn't the township make the developers pay extra to defray such costs, someone asks.

These questions touch on one of the conundrums faced by all communities in the path of development. The township is legally obliged to accept new residential subdivisions, but the extra property taxes generated by the new residents will not be sufficient to pay for the costs of the additional fire protection, policing, road maintenance, and garbage collection. More families also mean bigger schools. Townships that are next to major highways can hope to attract office buildings or shopping centers, which pay higher commercial property taxes and don't increase school enrollment or generate traffic on local roads. But there aren't any major highways going through Londonderry. That means that, in the long run, property and school taxes will probably go up.

Benner, unflappable, deftly steers the discussion from one topic to another without actually responding to the questions. He asks if someone from Arcadia would like to speak. "It makes sense to have my dad lead the conversation," Jason says later. "He has the most experience and the most stature." Joe Duckworth, with his longish hair—no beard this month—and casual clothes doesn't look like most people's idea of a developer. He describes New Daleville as a walkable community with lots of open space. Unlike Comitta, he steers clear of polemics. New Daleville is an alternative, he says, a different sort of housing. "It may not be for everybody," he adds. He politely reminds the audience that it was their neighbor who decided to sell his farm. Arcadia got involved because the township was unhappy with business as usual. He makes himself sound like a guest in their home. He also calmly but forcefully makes the point that the Wrigley tract will be developed, one way or another. The question is how.

Duckworth explains the master plan. He describes it as a compact village, with sidewalks and public greens. There will be walking trails and play lots. The garages will be behind the houses, he says, reached by rear lanes. He makes the point that half of the site will be left unbuilt. He also talks about the sewage treatment system, which will infiltrate treated wastewater into the ground. "Where will this take place?" asks a man sitting near the watercooler. Duckworth points to an area of open space. This raises a clamor, since it sounds as if the unbuilt area he had mentioned earlier is really a septic field. Duckworth explains that the sewage treatment area is not included in the open space calculation, but the misunderstanding leaves the impression that he has been caught cheating.

"We know what's in it for the developer," someone says brusquely, "but what's in it for us?" He doesn't mean the people who are going to live in the new houses—they are not represented in this room—he means the people who currently live in the township. Benner points out that the developers will donate land for a future township building and that the open space on the site will be for everyone. There appears to be skepticism in the audience on this point. Is it really going to be public? "It's up to you," says Duckworth. "We could deed it to an agricultural trust that would keep it as farmland, or give it to the township for recreational uses, ball fields and so on." This sounds too good to be true—a developer giving something away. "The proposed ordinance says that there may be up to twelve thousand five hundred square feet of retail or office space," says a woman who has obviously taken the trouble to read the document. "I'd hate to see golden arches here." "Or a Wal-Mart," someone chimes in. Duckworth assures them that 12,500 square feet is very small, not a large building, something like a convenience store or a professional office.

There are a number of questions about the houses that will be built in New Daleville. How big will they be? More important, how much will they cost? The latter is a key issue for the people who live in the nearby subdivisions. They own one-acre lots, whereas Arcadia is proposing something much smaller, and they are concerned that cheaper houses will bring down their property values. Duckworth says he expects the houses to be about two thousand square feet in size and to sell for about $200,000, which is comparable to current local house prices. He adds that, in other neotraditional developments, prices have generally risen over time. In that case, a woman asks, why does he need so many houses, why not a few less? "We're hoping that the prices will rise," says Duckworth, "but it's still risky." He explains that the larger number of houses is needed to cover the costs of curbs, sidewalks, lanes, and landscaping. He doesn't say that it also covers the cost of the long approval process, drawn out by meetings such as this one.

Despite Duckworth's well-considered attempt to persuade the audience, the general mood remains one of mistrust and antagonism. It's clear that if a floor vote were taken, a majority would vote against New Daleville. These people have not been impressed by neotraditional development. Walkability, the village concept, even the large amount of public open space, have not swayed opinion. As far as they are concerned, New Daleville could be as pretty as a Currier & Ives print but it's still a new development, that is, it's new houses on what was previously farmland. New houses mean extra cars, extra traffic at rush hour, more kids in the schools, and in the long run, higher taxes. Above all, new houses mean more people. The residents of Londonderry live here because they like the remoteness and the open countryside. They put up with driving some distance to work and to shop. Their isolation will be diminished by *any* development, whether on big lots or on small. Arcadia has a long way to go to convince them that New Daleville is a good idea.

The discussion lasts an hour. Citing a full agenda, Benner thanks the Arcadia team and calls for a brief recess. Jason and Della Porta, who have not spoken during the meeting, collect the presentation boards and go out to the parking lot. Nearby, the lights of Mindy Acres gleam prettily in the darkness. It's quiet, except for the low hum of occasional traffic on the highway, several miles away. The vast dome of the sky sparkles with stars. It's easy to see what attracts people to live here, and why they are reluctant to accept change.

Jason, who was hoping for a vote tonight, looks glum. "I thought that went well," Della Porta says to cheer him up. "These meetings can be much worse, with shouting and insults. Tonight was pretty mild." Comitta is not so sure. He's surprised both by the large turnout and by the hostile atmosphere. "The previous meetings of the planning commission rarely had more than three or four people in attendance. I've never heard so much negative comment about this project." He has to go back inside, since the preliminary plan of Honeycroft Village is slated for a vote. He doesn't expect any serious opposition. "Honeycroft is up in the northwest corner of the township, and it has few neighbors, so it's not in anyone's backyard," he explains. "In addition, the developer promised that the town houses would be mainly for retired and older home buyers without children, so schooling is not a big issue." He's right; the Honeycroft preliminary plan is speedily approved.

"In thirty years I've never been asked by the county to defend a proposed township ordinance," says Comitta, who will represent Londonderry before Chester County. He thinks the planning commission may be concerned that the county master plan designates Daleville as rural, whereas the new ordinance proposes a village. In fact, when Duckworth got involved in the Wrigley tract, he verified that the county was not opposed to a zoning change.

Duckworth believes that the unusual request for a formal presentation is caused by the fact that he himself is a member of the Chester County planning commission, and currently serves as chairman. "The commission staff are just being very careful," he says. "They want to make sure that there is no perception of favoritism or conflict of interest."

The county offices are located on the outskirts of West Chester, in a large, modern building that Comitta, when he gave me directions, described as resembling "a really big Circuit City." The meeting is in a spacious room on the first floor. There are five commissioners, three young women—a township supervisor, the president of a local resources council, and a foundation president—a young man who is a lawyer, and an older man who is a borough councilman. They sit at a U-shaped table. Comitta comes early with a stack of presentation boards, followed by Jason and Della Porta. Comitta will do most of the talking, but they are here in case there are any questions he can't answer. Joe Duckworth is the last to come in, very businesslike in a dark blue suit. He excuses himself for being late, explaining that he has been meeting with the county commissioners.

As chair of the commission, Duckworth sits at the head of the table and opens the meeting by calling for a minute of silence—it is September 11, a year after the attack on the World Trade Center. Approving the minutes of the last meeting, the commission launches into a flurry of motions, seconds, and votes on dozens of zoning ordinances and sewage facility plans. There is no debate, since the staff has already reviewed these cases, and the motions pass rapidly. Following a brief discussion of agricultural easements and land conservation, the agenda arrives at the Londonderry ordinance. Duckworth explains that, as the developer of New Daleville, he is recusing himself from the proceedings. He leaves the table and sits at the side of the room.

The first to speak about the Londonderry ordinance is Bill Fulton, who is the executive director of the commission, in charge of the professional staff. He gives some of the background on the Londonderry ordinance, explaining that originally the county proposed that Fagg's Manor should be the village center for the township. The supervisors felt this was an inappropriate location, so the village designation was removed from the county master plan. However, a village center is needed somewhere, and this now appears to be Daleville, he says. No problem there.

Comitta gets up to speak on behalf of the township. He hands out copies of the current zoning map and a short description of the neotraditional concept. He places plans of New Daleville and of the original Dilsheimer proposal on an easel and outlines the differences. He says that the township wants to use the extra open area as a public playing field. One of the commissioners asks if there is any opposition to the proposed zoning change. Comitta says that there is overall support. He doesn't mention the tied vote of the planning commission.

After answering some general questions from the commissioners, Comitta sits down. A young man speaks on behalf of the county planning staff. He has some technical questions about water supply and the width of the streets. He says the staff is concerned that the proposed ordinance may be challenged as spot zoning. Spot zoning means that an ordinance is so narrowly applied that it benefits the owner of a particular property to the detriment of his neighbors. Comitta tells me later that spot zoning is a murky concept which has no precise legal definition but is often cited by opponents of a particular zoning provision. In the case of New Daleville, the county appears to be uneasy that the new ordinance is being applied solely to the Wrigley tract. Although there are a few technicalities to be ironed out, this seems to be the main issue. The staff member concludes by recommending acceptance of the ordinance. Minor adjustments can be worked out with the township later, he says.

The commissioners have few questions. Unlike during the previous township meeting, there are no serious objections to the zoning amendment. The truth is that Arcadia's neotraditional project is just the sort of development the county has been encouraging. Jason, who has clearly done his homework, responds to some of the technical concerns, and Comitta embarks on a complicated rebuttal of the objection to spot zoning. He says that the township solicitor will review the ordinance and communicate with the county. I have the feeling that everyone is merely going through the motions. Duckworth was right, it is just the staff being careful. When the vote is taken, all five commissioners raise their hands in favor. New Daleville is still on track.

9

Scatteration

Every year, year in and year out, the American home-building industry produces between 1 million and 2 million new homes, four out of five of which are single-family houses.

When Tom Comitta spoke at the township meeting, he characterized traditional neighborhood development as an alternative to sprawl. Whatever their opinion of development, most people believe that sprawl is bad. Conservationists decry the loss of agricultural land; proponents of mass transit don't like spending more money on highway construction; environmentalists oppose continued dependence on fossil fuels; sociologists claim that low-density suburbs undermine community; urban planners see suburban sprawl as consuming resources that would be better spent on revitalizing inner cities; architects object to sprawl on aesthetic grounds; and, of course, opponents of development see sprawl as their chief enemy.

It is not so simple. For example, sprawl is often blamed for urban poverty, on the grounds that peripheral growth removes jobs from the inner city. Yet Anthony Downs, a Brookings Institution researcher and a longtime critic of sprawl, has found no significant relationship between sprawl and urban decline. "This was very surprising to me," he wrote, "and went against my belief that sprawl had contributed to concentrated poverty and therefore to urban decline."[1]

What about sprawl using up land? Most people in Londonderry would tell you that sprawl threatens farmland, but there is no evidence that a shortage of agricultural land is a serious national problem; in fact, during the last three decades of rampant suburbanization, food prices have dropped, not risen.[2] Environmentalists make sprawl sound like a voracious monster. Yet America is not running out of land. One researcher has calculated that to house the entire population of the United States at a low suburban density of one family per acre would require an area smaller than the state of Oregon.[3] Only about 5 percent of the United States landmass is currently urbanized—that is, occupied by buildings, roads, and parking lots—compared with 20 percent devoted to farming, more than 30 percent covered by forest. The balance—almost half—is wilderness.[4] Indeed, as unproductive farms are abandoned and rural people move to urban areas, wilderness has actually increased. "If preserving large ecosystems and wildlife habitat is your priority," wrote John Tierney in *The New York Times,* "better to concentrate people in suburbs and exurbs rather than scatter them in the remote countryside."[5]

Perhaps one reason for the confusion about sprawl is that there is no widely agreed-upon definition.[6] Some describe sprawl as a particular type of low-density growth, and others as a symptom of runaway development. And for some it is merely a temporary stage in the urbanization process. Late-nineteenth-century photographs of upper Manhattan show brownstones and apartment houses surrounded by open space—which looks like the sort of scattered development commonly associated with sprawl, yet in relatively short order, the empty spaces were filled in, and sprawl turned into city. Most people think they know sprawl when they see it. But do they? Los Angeles is popularly considered an example of sprawl, yet the population density of its built-up metropolitan area is actually greater than that of metropolitan New York.[7]

Likewise contrary to popular belief, Los Angeles is not a city of freeways; it has the fewest miles of freeway per capita of any American urbanized area (which is why its freeways are so congested).[8] The least dense metropolitan areas in the United States are not around the new cities of the South and West but surrounding older cities such as Detroit, Philadelphia, and Boston. Between 1982 and 1997, the urbanized areas of all three increased more than five times as quickly as their populations.[9] This reduction in population density is chiefly the result of home rule. All three cities are surrounded by small, independent municipalities which have enacted zoning that restricts growth by requiring large lots or by creating other obstacles to development. This, in turn, reduces density and pushes new construction farther and farther into previously rural areas.

The media commonly fuel misperceptions about sprawl. A 1995 cover story in *Newsweek* titled "Bye, Bye, Suburban Dream" described the growth of Phoenix in alarming terms: between 1950 and 1994, the area within city limits increased twenty-six-fold although the population grew only tenfold.[10] Obviously a case of sprawl—or is it? When a city expands by annexation, it acquires empty land, as well as unbuildable areas such as wetlands and mountain slopes. If one counts only the parts of metropolitan Phoenix that were actually urbanized in the fifteen years leading up to 1997, the area of metro Phoenix increased only half as quickly as its population; that is, metro Phoenix grew *denser*. Moreover, in 1997 the population density per urbanized square mile in Phoenix was greater than the metropolitan density of Chicago, Boston, and Philadelphia.[11]

Sprawl is often contrasted with dense downtowns, as if the choice were between living in a suburban rancher or an urban high-rise. However, according to the 1990 census, the densities of American suburbs and cities are not vastly different: the average gross population density of suburbs was 2,149 persons per square mile, that of cities was 2,813.[12] The explanation for this similarity is the nature of the American housing stock. As one might expect, the majority of suburban dwellings — almost three-quarters — are one- and two-story buildings. However, considerably more than half of city dwellings are also one- and two-story buildings. In fact, only 5 percent of city dwellings nationwide are in buildings of seven stories or more.

If American suburbs and cities are more similar than different, why does the specter of sprawl loom so large in the public's imagination? One reason is that sprawl is often equated with suburbanization. Virtually all postwar metropolitan growth in the United States has been suburban, but not all suburban growth, as Los Angeles and Phoenix demonstrate, is sprawl. As Downs points out, "Sprawl is not *any* form of suburban growth, but a *particular* form of it."[13] (He lists low densities, leapfrog development, and extreme political decentralization as some of the traits.) Another reason that sprawl appears pervasive is that the effects of growth can be so visible. Since coming to Philadelphia, my wife and I sometimes drive through Bucks County to a large flea market near Lambertville, New Jersey. It's as much a chance to get out in the country as to look at cracked teacups. Bucks County, roughly halfway between New York City and Philadelphia, used to be strictly a rural area, then a place for weekend retreats; now city people are moving here permanently, drawn by good schools and relatively inexpensive housing. Over the last ten years, the quiet country roads we take have become congested thoroughfares, and the picturesque fields have filled up with housing developments and discount malls. In fact, development in the county is generally concentrated, and large parts of the countryside remain open, but that is not the view we have from the road.

A lot of the new houses in Bucks County are the work of K. Hovnanian Homes, which has built more than 150,000 homes across the United States since the company was founded in 1959. According to Ara K. Hovnanian, who is president and CEO, "The challenge for home builders is to try and figure out the type of housing that will be demanded by buyers, and where the demand will occur geographically. The good news is that, over the long term, the size of the actual demand for new homes is entirely predictable."[14] The predictability he describes is the result of three conditions. The first is population growth. Thanks largely to immigration, the U.S. population has been increasing every year by more than 2 million persons. These people need somewhere to live. The second is steadily increasing prosperity. As people are better off, they want newer, better-equipped, and larger homes. The third is mobility. New jobs don't necessarily coincide with existing housing, and as people move—from cities to suburbs, from suburbs to rural areas, from one coast to the other—they, too, need places to live. As a result, every year, year in and year out, the American home-building industry produces between 1 million and 2 million new homes, four out of five of which are single-family houses. Add to this new workplaces, new shopping places, new entertainment places, new schools, new hospitals, and new roads tying them all together, and you have a Monopoly game in full play.

It's unsettling to live in a state of perpetual upheaval. That's probably why sprawl has become a whipping boy for so many of the things we don't like about modern life: traffic jams, overcrowding, instability, change itself.[15] George Galster, an urban economist at Wayne State University, describes sprawl as "the metaphor of choice for the shortcomings of the suburbs and the frustration of central cities . . . a conflation of ideology, experience, and effects."[16] I have a friend who has lived in Chester County for the last fifty years. He originally had an old house on a piece of land large enough that he could shoot rabbits without disturbing his neighbors. Over the years, he has seen the surrounding horse farms gradually replaced by residential subdivisions. Naturally, he grumbles about the influx of newcomers, the increased traffic, the noise, the slow disappearance of his bucolic surroundings. More than a decade ago, he subdivided his fifteen acres into three lots, selling two and building himself a new house on the third. In other words, in a small way, he became a real estate developer. But if I were to call him that, he would be outraged—sprawl is always somebody else's fault.

According to the *Oxford English Dictionary*, the word *sprawl* first appeared in print in 1955, in an article in the London *Times* that contained a disapproving reference to "great sprawl" at the city's periphery. Lewis Mumford referred to "sprawling suburbia" in his 1961 classic, *The City in History*.[17] A 1965 article in *Land Economics* defined sprawl as "areas of essentially urban character at the urban fringe but which are scattered or strung-out, or surrounded by . . . underdeveloped sites or agricultural uses."[18] At that time, the more neutral term *scatteration* was also used to describe this phenomenon.[19] Thanks to a famous 1974 study titled *The Costs of Sprawl*, which computed the direct costs and adverse environmental effects of low-density development, *sprawl* entered the planning lexicon.[20] The methodology of the study was later called into question, but the term stuck.[21] There is no better way to occupy the high ground in a debate than to define its language.

The *Costs of Sprawl* study was prompted by the fact that in 1970, for the first time, more Americans lived in suburbs than in rural areas or cities. The authors of the study predicted that suburbanization between 1970 and 2000 would be almost as great as in the previous twenty years, which had been "the period of greatest suburban growth in the nation's history."[22] They underestimated on two counts. Suburban growth was greater than expected, not 70 percent but 80 percent, and the overall population grew not by 46 million but by 76 million. As a result, the increase in the number of people living in the suburbs turned out to be almost twice as great as predicted. The United States had become, in the words of one commentator, a "nation of suburbs."[23]

When railroads and streetcars opened up the urban periphery in the nineteenth century, only the well-off could afford to commute, whether it was from Chestnut Hill to Center City Philadelphia, from Brookline to downtown Boston, from Lake Forest to Chicago's Loop, or from Tuxedo Park to Manhattan. That might have remained the pattern—a select number of wealthy garden suburbs on the distant fringes of dense, blue-collar, industrial cities—but for Henry Ford. Inexpensive automobiles gave mobility to everyone.

John Nolen, who was a student of Fredrick Law Olmsted, Jr., and one of the most prolific American planners of the early twentieth century, predicted the revolutionary impact that cars would have on urbanization. In 1927 he wrote: "If the movement away from the cities assumes the formidable aspect of a hegira (and the magnitude of recent modern developments like the automobile and the radio makes this appear quite likely), then it is immensely important that it be organized and directed accordingly." He understood that technology was making suburban living convenient and attractive. "This sort of urban overflow is like that of a rising flood braking a dike at its weakest point and spreading uncontrolled over the adjacent country, greatly to its damage."[24]

Nolen's solution to suburban growth was to channel the overflow into planned garden suburbs, among them his exquisitely planned model town of Mariemont, outside Cincinnati. He believed in design, but unlike most city planners today, he was not wedded to high-density development. He agreed with his friend Raymond Unwin, who once wrote a pamphlet titled "Nothing Gained by Overcrowding!" Nolen and Unwin decried the congested tenements and walk-ups of the old industrial cities. They wanted everyone—not just the rich—to have their own homes, their own gardens, and access to nearby parks and playgrounds. Garden suburbs delivered that promise. Nolen and Unwin's suburban strategy still appears sound. As Gregg Easterbrook wrote in *The New Republic* in 1999: "If suburbs are where Americans choose to live—and that verdict is in, the suburban class now constituting the majority of Americans—then brainpower should be applied to making burbs as livable as possible."[25]

One of the planning ideas advanced as an antidote to scattered development is so-called smart growth, which originated in the nineteen nineties.[26] Smart growth, like sprawl, is a slippery concept, not least because it is espoused by anti-growth environmentalists as well as pro-growth developers. According to Anthony Downs, advocates of smart growth do have some things in common. They are for walkable communities and mixed-use town centers, and generally favor preserving open space and redeveloping inner cities. However, depending on who is speaking, smart growth can also include controversial ideas such as subsidizing mass transit to reduce car dependency, creating regional governments, and establishing urban growth boundaries to restrict growth into rural areas. While environmentalists see smart growth as a way of placing limits on growth, developers would like to change zoning to permit higher densities, and land conservationists would like to restrict development to selected areas. Downs concludes that, as a national strategy, smart growth is simply too contradictory to be effective, and he argues for elements of smart growth to be applied selectively at the regional level. As he succinctly puts it, "What is 'smart' in New York City may be 'dumb' in Phoenix."[27]

So, what is smart in Londonderry? On the one hand, if sprawl is measured in consumption of land, New Daleville, with more lots on less space, appears to limit sprawl. Compared with the 86 houses that were originally planned for the Wrigley tract, New Daleville will have 125, which is an increase of almost 50 percent. However, since the lots at New Daleville will be smaller, it is likely that the houses will appeal to smaller families and empty-nesters. If the average family size in New Daleville is three rather than four, the total population will be 375 persons, versus 344. Still an increase, but nowhere near as dramatic.

On the other hand, if sprawl means building over farmland, then New Daleville will contribute to sprawl. Since Londonderry has no real master plan, merely zoning districts, the development, however well-designed, will remain an isolated residential island, just like Mindy Acres across the road. Although Bob Heuser has designed a walkable community, there is not really anywhere to walk to, since there is no real village center, nor is there likely to be. Since the density of Londonderry will always be too low for mass transit, the future inhabitants of New Daleville will be heavily dependent on their cars. Their comings and goings will add to the traffic and congestion of the back roads of Chester County—according to most estimates, New Daleville will generate more than a hundred peak-hour car trips daily. Thus, for hard-core, transit-first, rebuild-the-center-city, regional planning advocates of smart growth, New Daleville is merely more of the same, what they don't want.

Yet New Daleville's compact layout will likely foster a greater sense of community than if the houses were spread out. Children will play in the parks—and probably in the lanes. People will more easily meet their neighbors. They may even organize public events on the common green. With its compact plan, New Daleville will be a nice place to walk—for exercise and for pleasure. The narrower streets and denser layout will reduce the amount of asphalt, hence there will be less polluted runoff; more rainfall will be absorbed into the ground naturally. Half of the site will be left unbuilt in perpetuity—no small accomplishment. Kids will be able to walk or bicycle to the playing fields. Above all, New Daleville, unlike subdivisions in the area, will include shared, public spaces: sidewalks, walking trails, play lots, village greens, parks. These will be small reminders to the people living there that they are not only private homeowners but also members of a community. That will be smarter growth indeed.

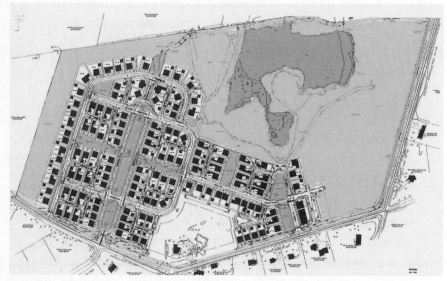

New Daleville Plan, October 2002

10

More Meetings

After upzoning here and downzoning there, the township passes an ordinance.

The Londonderry supervisors don't meet again for a month. Meanwhile, the plan of New Daleville continues to evolve. The Wrigley homestead, surrounded by five acres, divides the site into two major parts. The planner, Bob Heuser, has located the park and the village center on one side, and most of the homes on the other. "I've organized the plan around two main axes," he says. "The boulevard runs north-south, and the street that links that neighborhood to the village center goes east-west." The general location of the village center is dictated by Joe Duckworth's demand that it be visible from the main road. The center has sites for a church, commercial building, and a village green. Most of the house lots are about one-eighth of an acre. Two dozen larger lots at the edge of the site—Duckworth's ocean view lots—have driveways; the rest of the houses are served by rear lanes, a standard neotraditional arrangement.

Like all developers, Duckworth keeps an eye on what his competitors are doing. He learned this while he was working at the home builder Toll Brothers. At that time, he was in charge of a project in a part of Bucks County where a small developer was having great success with a Mediterranean-style model that seemed to appeal to Hispanic buyers. The house was stuccoed, with Spanish Colonial features. "Bob Toll said that we had to build some of those," remembers Duckworth. So one night he sent a foreman to the competitor's development to measure one of the houses. "You see, we didn't know exactly what it was that appealed to people, so we simply copied the whole thing."

Duckworth has recently seen an interesting new type of house group in Denver, at Stapleton, a large neotraditional development on the site of an old airport. The houses are clustered around common courts, set at a right angle to the street. The intimate green courts make safe and secure play areas for children, similar to cul-de-sacs but without the cars and the asphalt. Housing courts are not new—groups of houses around common greens were built in Chestnut Hill in the nineteen twenties—but what catches Duckworth's attention in Denver is the fact that, even though you can't park your car at your front door, the houses in these courts are actually selling more quickly than conventional street-facing houses. He describes the courts to Heuser and tells him to include some in New Daleville.

The village center gives Heuser the most trouble. Since there are so few public buildings, it isn't easy to create the villagelike atmosphere Duckworth wants. Early on it is apparent that the township is not interested in a site for a church. Instead, it wants land set aside for some sort of civic building, perhaps a firehouse or a new township hall. To show goodwill, Duckworth donates the required land to the township.

The entire Arcadia crew—Joe and Jason Duckworth, Dave Della Porta, and Bob Heuser—show up for the supervisors' October meeting, when they are due to vote on the New Daleville ordinance. Tom Comitta knows that some of the supervisors have been rattled by the county's questions about spot zoning but hopes this issue has been laid to rest. "The one thing about this whole science of zoning and planning is that it doesn't necessarily follow rules, as you might expect," he tells me. "It follows state legislation, and it somewhat follows logic, but zoning boards can be very arbitrary. Often the decisions are unusually subjective. One board will vote one way, another board, on the same question, will vote another way."

The room is overflowing; more people have come than even the previous time. They are here not for New Daleville, however, but for the first item on the agenda, a sweeping proposal to rezone large parts of the entire township. Tim Cassidy has explained to me that the rezoning was initiated by the large estate owners. With the help of the Brandywine Conservancy, they approached the Londonderry supervisors with a proposal to create what is called an "agricultural preservation zone." The chief characteristic of this zone will be that all new residential lots must be at least twenty-five acres. This will stop conventional real estate development in its tracks. The conservancy has prepared a study of soils, slopes, and wetlands to bolster its case, and has also submitted a petition signed by fully 80 percent of the landowners. The proposed preservation zone affects two-thirds of Londonderry—almost five thousand acres. The supervisors have agreed to consider the change. Since Honeycroft Village and New Daleville will increase the density in some parts of the township, they feel that the density of other residential districts can safely be decreased. As Cassidy succinctly puts it: "If we upzone here, we can downzone there."*

Howard Benner announces that the supervisors will vote on the proposed downzoning later that evening and opens the meeting to public discussion. A man in the front jumps up. He isn't happy with the proposed change. He says that he owns sixty acres immediately north of the New Daleville tract. He bought this land fifteen years ago and is currently negotiating with a builder. His land is in the downzoned district, so now only lots of two acres

*Upzoning generally means upping—or increasing—density, although in the South the sense is reversed and upzoning means upping the restriction, that is, reducing density.

will be allowed, not one acre, as before. In other words, his land will lose half its development value if the change goes through. After reading his prepared statement, he asks, "If the change is approved, can I still develop my land according to the previous zoning?" Bob Harsch, the township engineer, answers that no, once the zoning is changed, the new rules will apply to everyone. The landowner, previously calm, grows angry. He says that he is being unfairly targeted, since three-quarters of the land that will be downzoned belongs to him. Having vented his anger on the supervisors, he turns to the audience. "Your property values have just gone down by fifty percent," he says. "How can you stand for it?" He quotes a passage from Deuteronomy about the sanctity of ownership. He says that downzoning is un-American. It is nothing short of terrorism, he says. He seems to be speaking in sound bites, but there are no media to hear him, only a roomful of embarrassed neighbors. Finally, he runs out of steam. "It's just not right," he says lamely and sits down.

The next question is from a teenage boy who has come in late with a group of friends, all carrying skateboards. "Why do we have to have new houses here anyway?" he asks, somewhat belligerently. It's a question that must be on a lot of people's minds. Benner patiently explains that no doubt the majority of people in Londonderry would be happiest if there was no new development. "But we can't just stop it," he says. "The land has been zoned to allow housing."

A man in a suit stands. He is a lawyer representing another property owner whose land has been downzoned. He says that his client is asking not for a denser classification, only that the existing, one-acre zoning stay in place. After the Bible-quoting landowner, he sounds reasonable, and his argument is compelling. Moreover, it turns out that his client submitted a subdivision plan for approval just before the meeting, so it is likely that he will not be held to the new zoning. But the lawyer is making sure the supervisors understand the situation. "He's throwing himself on the mercy of the board, plus he's actually made a convincing case," Duckworth whispers to me. "I would be surprised if the final rezoning doesn't accommodate him."

There are no more questions about the rezoning, and Benner invites Comitta to present the New Daleville ordinance. Comitta describes what happened at the county planning commission meeting. He lists the changes recommended by the commission staff. They are all minor items, he says, but he goes through them somewhat pedantically, point by point. He makes it sound both routine and boring. He pins up a plan of New Daleville and reminds the audience that, while there are more houses, half the site will be open space and will belong to the township. He closes by saying that, since the state passed enabling legislation for neo-traditional development two years earlier, about two dozen townships have adopted such ordinances. These types of developments are a known quantity, he says. He refers to them as "Ozzie and Harriet neighborhoods."

One of the boys with skateboards asks if the open space could include a skate park. "There's a skate park in the area, but our parents work and it's too far for us to go," he says. Comitta answers that it is up to the township to decide what it wants. "You should give them your ideas," he says. "Make sure you come to the next meeting." This appears to satisfy the boys. Their business concluded, they pick up their skateboards and leave. There are no other questions.

Benner calls a recess so the supervisors can meet with the township solicitor and the engineer. He doesn't explain the reason for the private consultation. They go to the adjacent township office and close the door. (Pennsylvania Supreme Court case law allows for executive sessions in zoning and land-use matters.) The Arcadia people stand together around the watercooler in the back of the room. There has been no call for anyone to speak. Basically, things have gone well, but they still look nervous as they wait for the meeting to resume.

Five minutes later, the supervisors file back in. Benner calls the meeting to order and asks for a motion on the text of the proposed townshipwide downzoning, which passes unanimously. He says that the supervisors have decided not to vote on the zoning map itself until after it has been advertised and more fully discussed in public. He announces that the planning commission will meet the following week to talk about the changes to the zoning map. Then he calls for a vote on the New Daleville ordinance. "All in favor say aye," he instructs, and he, Clair Burkhart, and Fred Muller do. Londonderry has a new ordinance, and Arcadia has a project.

"Now we can start," says Duckworth after the meeting, obviously relieved. The new ordinance is an important legal hurdle, since without it New Daleville could not exist. But the ordinance does not guarantee that the project itself will be approved; that's the next step. New Daleville is on schedule. Seven months have passed since Duckworth was approached by Dick Dilsheimer. Now Duckworth can give the go-ahead to the consultants to start working on the engineering plans. If all goes well, New Daleville will receive preliminary approval a year from now, and final approval several months after that. It looks as if they will break ground in early 2004, as planned.

"I feel fairly comfortable about the project. The biggest issues are behind us," says Duckworth. Although Comitta is concerned that the new ordinance has been publicly linked to the controversial township rezoning, so a lawsuit against Londonderry by a disaffected landowner could tie up the New Daleville ordinance, Duckworth is not worried. "We shouldn't be affected by anything that irate citizens decide to do legally."

His general optimism is buoyed by a recent announcement that, despite rising prices, the annual construction of housing units in the United States is heading toward 2 million, and the national home ownership rate now stands at an all-time high of almost 70 percent.[1] Low interest rates are encouraging home buying. Even if interest rates rise, Duckworth believes that demand for housing in Chester County will remain strong because of an undersupply of permitted land. In any case, for the moment Arcadia doesn't have much financial exposure. "We can walk away now and we're out a couple of hundred thousand. It's only after we close on the land, early next year, that it will be different. Then we'll really be committed," he says.

Duckworth plans to use as little of his own money as possible to buy the land. "What I learned from Bob Toll, and at Realen, was how to tie up large investments in land with small amounts of cash," he says. "You can usually count on prices going up because of pent-up demand. So when it's time to go to a bank, you can say, 'Look, this is worth a lot more now than it was before.' And if I can show the bank that I've put in my own equity, and I have a builder lined up, the bank might even lend me a hundred percent of what I need. There's an old real estate saying: A dollar borrowed is a dollar earned, a dollar paid back is lost forever."

PART TWO

New Daleville, April 2005

11

Drop by Drop

To a developer, sewage disposal is one of the most complicated and expensive steps in improving raw land. Arcadia has budgeted $1.5 million for the construction of the treatment and disposal facilities, the largest single expense other than the cost of the land itself.

The Arcadia offices are in a small but imposing granite building at the main crossroads of Wayne, a suburban town on Philadelphia's Main Line. WAYNE TITLE AND TRUST COMPANY carved across the fascia is a reminder of its original function. From the front door, you can see a collection of shops and restaurants, a hotel, a cinema, and two church spires. The commuter railroad station is around the corner. When the Duckworths want to demonstrate the benefits of a walkable village center to their clients, they don't have to take them far.

Arcadia occupies part of the ground floor, three private offices around an open space. A round conference table is piled high with papers. The office, which seemed large enough a few years ago, is overflowing with filing cabinets, presentation boards, and stacks of reports. Jason is looking into leasing the space next door. He's been working with his father for a little more than a year, and he wants to expand the operation, do more projects, and take on more staff. Joe says that's fine with him, but he has two conditions:

101

the only person to report to him will be Jason, and any growth in the business has to pay for itself.

It's April 2003, six months since Londonderry approved the neotraditional ordinance, and New Daleville seems to be moving ahead smoothly. Today Jason is at his desk, telling me about the engineering drawings. These documents are the most expensive part of the entitlement process. The forty-one sheets describe the layout of the lots, the grading of the streets and lanes, the detailed design of water lines, sewers, utilities, storm water collection, and landscaping. Arcadia has budgeted $150,000 for this work, but it looks as if just the preliminary phase will cost $90,000. The grading has proved more complicated than anticipated, since the site slopes and the rear lanes, which have to be at the correct height to line up with the garages at the backs of the lots, also have to line up with the streets.

Jason tells me that, the previous evening, he and Dave Della Porta attended a meeting of the Londonderry planning commission to explain the latest version of the plan. The room was packed with more than forty people, and Jason was concerned at the size of the crowd. It turned out that most were there to voice their opposition to an application to build mushroom growing houses in Londonderry. Although Chester County produces 40 percent of all mushrooms grown in the United States, mushroom houses are unpopular with local residents, who objected to the smell of compost, the flies, the spores, chemicals, and pesticides, as well as the truck traffic and the presence of migrant laborers. By the time Jason got up to speak, New Daleville sounded downright benign.

Jason tells me that he is worried about one thing. Although he filed the engineering plans with the township several weeks ago, he still doesn't have the final planning approval of the sewage treatment system, or the sewage planning module, to give it its technical title. The township says it will not vote on the preliminary plans without the module. Jason has been stalling, but the truth is that the sewage plan is far from ready.

Sewage is not something most people give a lot of thought to; it's out of sight and out of mind. But to a developer, sewage disposal is one of the most complicated—and expensive—steps in "improving" raw land. Arcadia has budgeted $1.5 million for the construction of the sewage treatment and disposal facilities, the largest single expense other than the cost of the land itself. Planning the system is the responsibility of Jeff Miller. His consulting company, Evans Mill Environmental, is located above a post office in a nondescript roadside commercial building in northern Chester County. Miller is a burly, bearded man in his fifties who has a master's degree in resource management and environmental engineering and looks as if he spends a lot of time outdoors. Evans Mill deals with wells, environmental remediation, and small-scale wastewater treatment. "We also do endangered species," Miller tells me as we talk in his cluttered office. "Around here that's mainly bog turtles, which seem to be everywhere, which is odd considering they're endangered."

Miller explains the arcane subject of suburban sewage treatment. In the absence of a municipal system, residential developers usually build individual septic tanks, which require lots of at least an acre. A development with smaller lots, such as New Daleville, needs a so-called community system, that is, a network of sewer pipes leading to a treatment facility. The usual technique of disposing of treated wastewater is to spray it on the surface in a restricted area and let it sink into the ground. The alternative, which is what Miller is proposing for New Daleville, is called drip irrigation. The treated wastewater is pumped underground through perforated tubes that are placed directly in the soil, in shallow slit trenches, cut by the same sort of machine that lays underground fiber-optic cables. Miller shows me a sample of ordinary-looking plastic pipe, half an inch in diameter, with punched perforations. The key to the system is the design of the perforations, he explains, which are shaped so that, whatever the pressure in the pipe, the water drips into the ground at a constant rate.

Drip irrigation, which is based on agricultural technology developed in Israel, was introduced to the United States in the late nineteen eighties. It costs more than spray irrigation, but according to Miller, it has several advantages. Spraying creates aerosols that can be transported by wind, so the spraying area has to be far from humans. Underground drip fields can be as close as twelve feet to houses, driveways, and streets; moreover, the surface of a drip field can be used as a playing field or a park.

Miller still hasn't finished the sewage planning module for New Daleville. The problem is the soil testing. The tests are crucial, since he needs to know the precise permeability of the soil, as well as the depth of the water table, before he can design the system. According to Pennsylvania environmental regulations, the saturated soil below the drip pipe, called a mound, has to be at least four feet above the water table. That means the shallower the water table, the smaller the mound, the smaller the mound, the slower the rate of flow, and the slower the flow rate, the larger the area of the drip field. Permeability is measured by digging a hole to the required depth, placing a plastic pipe in the hole, sealing the space around the outside of the pipe, and measuring the rate at which water poured into the pipe is absorbed. The depth of the water table is ascertained simply by drilling a deep hole and measuring the distance to standing water. Both tests have to be done in the presence of a soil scientist from the Pennsylvania Department of Environmental Protection, or DEP.

As soon as the New Daleville ordinance was passed, in October, Miller called the regional office of DEP to make an appointment for carrying out the tests. He was told to call back in January, since the soil scientist was booked solid until the end of November and was on vacation during December. After complaining, Miller finally got an appointment in mid-January. It was a particularly cold winter, and by January the ground was frozen solid. The department requires that tests take place only after the soil has thawed, which did not happen until late March. The water-table test was delayed even longer. Because of a late spring thaw, the drilling rig could not be moved into the muddy field until May. The normal procedure is to drill holes deep into the aquifer, then wait for several weeks until the groundwater level settles to predrilling levels. The holes were drilled, but since it rained all month, they kept filling up with rainwater. It now looks as if the results will not be available before early June, a total delay of six months.

Miller is not worried. A year ago, when Duckworth first heard about the Wrigley tract, he asked Evans Mill to make a preliminary check of the water-table depth. "We dug seven-foot-deep pits," Miller says. "We didn't see any water. Of course, we were coming off a drought, but we weren't worried. Typically, in this area, you get a deep water table and good permeability. In a bad situation, you might get a shallow water table with good permeability, or vice versa. In either case, you can adjust the rate of flow to compensate."

When the results of the soil tests finally arrive, the news is unexpected. "It couldn't be worse," says Miller. "What we have is a rare combination of poor permeability *and* a shallow water table." To accommodate these conditions, he has to reduce the rate of flow drastically, which means increasing the size of the drip field, from seven and a half to twelve acres. The problem is that there is not enough room to do this. In other words, New Daleville has hit a major snag.

"We have several options," says Duckworth, "none of them good." Miller has suggested he could try to convince DEP to accept a lower figure for household wastewater production, although he acknowledges that this is a long shot. Or he could ask Chester County to allow some of the treated wastewater to be discharged directly into a stream. "Jeff says that stream discharge is environmentally okay, since the wastewater has been subjected to tertiary treatment, which means it's completely safe. But the policy of the county is to encourage on-site disposal, so this too is a long shot," Duckworth tells me. Another option is to enlarge the drip field by buying a piece of land from a neighbor. "We might need as much as four or five acres," he says. "That would not only be expensive, there's no guarantee we could find a seller." The least attractive option, from Duckworth's point of view, is to replan New Daleville with fewer lots. He estimates that he would have to reduce the lots to fewer than a hundred, which means the project would barely break even.

Although Duckworth has not yet bought the land, he is not ready to abandon the project. "We've already invested time and money, so we would definitely prefer to develop the land," he says. "We're trying to figure out how to make it work." He is philosophical. "It's nobody's fault. There's a higher risk working with new technologies. But there's no doubt that this is a major setback, especially since the New Daleville project has had the simplest entitlement process I've seen in years. This is typical. If it's not one thing, it's another." Duckworth cultivates an upbeat image, but I can see that he's disappointed.

Miller suggests another possible solution. He thinks that the strip of land—about ten acres—the township insisted be left as a buffer next to Charlotte Wrigley's farm, on the other side of the site, might be large enough for a drip field, depending on the permeability of the soil and the height of the water table. He applies for a permit to drill half a dozen new test holes in that area. A few weeks later, the results of the new soil tests come in. It turns out that, while the ground in the new location is of similarly slow permeability, the water table is deeper—forty-six feet instead of thirty. The extra depth allows Miller to increase the flow rate.

There is one problem. Because of the geometry of how the underground pipes are laid, Miller needs about twelve acres for the new field, which means widening the strip of land. Bob Heuser will have to modify his plan. After examining several alternatives, he finds a way to provide more space without sacrificing a single lot. He does this by moving one street and creating a small cul-de-sac. The new plan not only has more premium lots backing onto green space but actually reduces the amount of roadway. The planner has saved the day.

With this final change, Heuser's work on New Daleville is finished. During the last eighteen months, he has produced a variety of drawings. The early conceptual plans were drawn at a small scale (one inch equals one hundred feet), but the final plan, which he called a "pre-engineered site plan," is drawn at a large scale (one inch equals fifty feet) and shows houses and garages as well as details such as sidewalks and curb cuts. "This drawing can be scanned by the engineers and used as a base plan for their technical studies," he explains. "I make it very accurate to make sure that the engineers do engineering, not planning." With all the changes, his fee has grown to about $20,000. Although Heuser is in many ways the creator of New Daleville, this is probably among the smallest of all the consultants' fees.

Jason, who is assuming more and more responsibility for the project, is relieved by the outcome. "I wasn't sure that we would have a solution," he says. It's now August, and he intends to submit a revised preliminary plan to the planning commission in early September. "I've learned a lot more about soil and geology than I expected to," he adds ruefully. Before the sewage planning module can be sent to DEP for final approval, it has to be reviewed by the county. Once the county comments come back, a public notice of the sewage plan will be posted for thirty days, which will give interested citizens a chance to review the plan at the township offices.

The revisions to the plan are received well by the township. The planning commissioners had originally requested the buffer areas on the north and west site boundaries, so they are pleased that these are now slightly wider. They like the location of the treatment plant, which has been moved to a low part of the site, behind the woods, and is less visible than before. They also appreciate the fact that the sewage disposal field is no longer beneath the park. Most of the public discussion at the meeting centers on how the park will be used. Tom Comitta's office has prepared a plan for a ten-acre neighborhood park, including a soccer field, tennis courts, a volleyball area, a children's playground, and a small garden pavilion. These facilities share the parking lot of the New Daleville village center. But there is no agreement among the planning commission members about exactly what sort of recreation space is needed. Should it be baseball or soccer? Should they do something now or wait until they have a comprehensive recreation plan for the entire township? Some of the commissioners think that elaborate recreational facilities are unnecessary. They played pickup ball in mown fields as children, why does everything have to be organized? Times have changed, another of the commissioners points out, our kids expect equipment and a proper field, and in any case there are legal liabilities to consider. Underlying the discussion is the question of money—Comitta estimates that the park will cost almost a million dollars. The commission hopes the developer will pay for it.

Arcadia has no intention of covering the entire cost of a park. "We'll pay anything within reason, if it means that the approval process will go faster," says Jason. "The problem is that the township doesn't seem to know what it wants. There is already a soccer field next to the Londonderry Township building, but they still haven't bought the goalposts. So we're a bit skeptical. Playing fields or farm fields are both okay with us. We would just like to get a clear answer from them soon."

A year has gone by since the New Daleville ordinance was approved. Because of the problems with sewage treatment, the project is now more than nine months behind schedule, and Arcadia still doesn't have preliminary approval for its plans. In the meantime, from Jason's point of view, the meetings with the township have not been beneficial. Between Comitta's office and the planning commission, he is getting too much advice. The village center has gone through several modifications, because the township keeps changing the details of the playing field and also can't decide if it really wants a site for a new township building. At the end of every meeting, Jason has a checklist of changes and additions. These are incorporated into the plan, but at the next meeting new issues emerge: the need for hedgerows, the size of street trees, the illumination levels of street lighting, and the surface of the walking trails. It seems as if it will never end.

12

On the Way to Exurbia

*For the first time in history, urbanization does not mean
concentration.*

Despite Arcadia's claims, New Daleville will not really be a
village, and it is too rural to be called a suburb; it's what city
planners call an exurb.[1] Exurbs are a relatively new phenomenon.
Nineteenth-century garden suburbs such as Riverside and Chestnut Hill were far from the city but were firmly tied to downtown
by railroads and streetcars. The first postwar suburbs were called
bedroom communities precisely because their inhabitants only
slept there but worked, shopped, and played in the city. Exurbs
are different. "They have broken free of the gravitational pull of
the cities," writes David Brooks, "and now exist in their own
world far beyond."[2] Such extreme decentralization is the result
not simply of car ownership but of mobile communications, pri-
vatized entertainment, and an increasingly effective distribution
of goods and services. These technologies have enabled people to
work as well as live far beyond the urban periphery, in areas that
were previously entirely rural, such as Chester County. For the
first time in history, urbanization does not mean concentration.

This dispersal is apparent in New Daleville, where everyday life
will be marked by an extreme and far-flung mobility. The future
inhabitants will drive to work in West Chester and Downingtown,

shop at the Exton Mall, and buy fresh vegetables from Amish farm stands in Lancaster County. For Christmas gifts they may go to King of Prussia. If they get ill they'll be taken to the Southern Chester County Medical Center, about three miles away. Presbyterians will go to church in Fagg's Manor, Baptists will travel to nearby Cochranville, and Roman Catholics will attend Our Lady of Rectory in Parkesburg. School buses will take the kids to the Octorara educational campus, six miles away. Mothers will ferry their kids to soccer practice, and to Chuck E. Cheese's and the multiplex on Saturday afternoons. They may join one of several golf clubs in the area. For recreation they'll go to Longwood Gardens or the Brandywine River Museum. They will occasionally visit Wilmington or Philadelphia, but for most of them the visits will be few and far between.

Exurbia is recent, but its advent was foreseen a long time ago. In his 1902 book about the future, *Anticipations,* H. G. Wells titled a chapter "The Probable Diffusion of Great Cities."[3] He observed that the way people live is always a function of transportation. A city of pedestrians is limited by a radius of about four miles; a horse-using city by seven or eight; but the radius of a city with suburban trains could easily expand to thirty miles.[4] Wells predicted that faster trains, omnibuses, telegraphs, telephones, and something he called "parcel-delivery tubes" would make the concentrated city obsolete. He imagined well-off city people living far out in the countryside. He was vague about the physical form of what he called "urban regions," except to say that they would be extremely spread out and would definitely not resemble traditional cities.

It was left to another visionary to flesh out this idea. Frank Lloyd Wright had spent most of his early working life in and around Chicago, but it was not that city that stimulated his ideas about the future. In 1922, after a decade spent mainly in Europe and Japan, the already-famous architect settled in Los Angeles.[5] Los Angeles in the twenties was the fastest growing city in the country, but it was growing in unusual ways. Despite its relatively small population of half a million, the city stretched from San Bernardino to Santa Monica, a distance of nearly seventy miles. The Pacific Electric Railway, which operated a thousand miles of track, provided access to this vast region, as did private automobiles. With the highest rate of car ownership in the world, and a network of boulevards, drives, and highways, Los Angeles was unlike any large city in the United States—or anywhere else.

The chief reason for Wright's move to California was to find new clients. In 1923, in concert with a local real estate developer, he designed a planned community on a four-hundred-acre tract in Beverly Hills. The speculative project was aimed at winning the financial support of the celebrated oil millionaire Edward L. Doheny, owner of the land.[6] Large-scale planned communities were nothing new in Los Angeles, and in fact the largest of these, Palos Verdes Estates, was under way at the time of Wright's proposal. Palos Verdes, developed by Frank A. Vanderlip, a New York City banker, was a classic garden suburb with one exception: it was intended for people with cars.[7] To preserve the spectacular natural beauty of the coastal peninsula immediately south of the city, its planners, Frederick Law Olmsted, Jr., and Charles Cheney, left half of the 3,200 acres unbuilt, in the form of nature preserves, parks, and golf courses. They dispersed schools, village centers, and golf courses across the entire site. Houses—in compact groups—were likewise scattered, connected by a far-ranging network of roads, including a twenty-mile scenic drive that was brilliantly fitted to the hilly coastal terrain. To make Palos Verdes "an ideal garden suburb and residence park," main thoroughfares for heavy traffic were kept apart from residential streets, which were designed to discourage fast driving.

Following Olmsted's lead, Wright adapted his plan to the topography of Beverly Hills and took advantage of automobile access to spread the houses across the steep terrain.[8] Unlike the traditionally styled residences at Palos Verdes, however, Wright's houses were decidedly original—Mayan/Art Deco compositions with hillside terraces and hanging gardens, and whereas Olmsted laid out the roads discreetly hugging the slopes, Wright used retaining walls and bridges to integrate the roadways dramatically with the architecture. The proposal was original and exciting, but there is no evidence that Doheny showed any interest.[9]

The Depression obliged Wright to close his West Coast office and return to Wisconsin, yet the influence of Los Angeles was not lost on him. In 1931 he gave a series of lectures at Princeton University, in which he argued that personal mobility was radically transforming traditional urbanism. The impetus to put his ideas on paper came soon after. In January 1932, *The New York Times Magazine* published an article titled "A Noted Architect Dissects Our Cities."[10] The author was the French-Swiss architect Le Corbusier, who described his radical vision of the modern city. "We must immediately discard the traditional type of house and allot to each inhabitant a soundproof living room, with plenty of light." These living rooms, as the accompanying illustrations showed, were in apartments in immense high-rise buildings set in parklike surroundings. Cars would be carried on elevated highways. "Everything will be new," he wrote.[11]

Wright despised Le Corbusier and the newly named International Style of architecture he personified.[12] Moreover he heartily disagreed with Le Corbusier's analysis. Only two months later, the *Times* published Wright's withering response.[13] What was the point of a skyscraper city? he asked. "Super space making for rent, to enable super-landlords to have and to hold the super-millions in super-concentration to make super-millions of superfluous millions?" Like Wells, Wright argued that decentralization—in transportation, communication, technology—was the preeminent modern trend. Instead of a vertical, concentrated city, he described "the horizontal line of the machine age, indefinitely extended as the great architecture highway and by the flat plane of the machine age expanded into the free acreage of the Broadacre

City."[14] Instead of bringing nature into the city, Wright proposed extending the city into the country.

Later that year Wright published *The Disappearing City,* whose dramatic title summarized his argument. His description of Broadacre City sounds like a vast version of Los Angeles, but without the downtown: "Giant roads . . . separate and unite the series of diversified units, the farm units, the factory units, the roadside markets, the garden schools, the dwelling places (each on its acre of individually adorned and cultivated ground), the places for pleasure and leisure."[15] This vision was based on a simple yet radical insight: "What we need is the wedding of the city and the country. . . . There you have all the advantages of the city, without the city."[16] Always an iconoclast, Wright later proclaimed, "In the City of Yesterday ground space is still reckoned by the square foot. In the City of Tomorrow ground space will be reckoned by the acre."[17]

Wright was always careful to stress that his ideas were not utopian. "It is nothing that I have invented," he once said.[18] "There is plenty of evidence now at hand to substantiate all the changes I outline."[19] He anticipated the coming of shopping centers, which he called "roadside markets," office campuses, suburban department stores, and service station convenience stores. Wright was interested in transportation and predicted the replacement of long-distance trains, then at their zenith, by air travel. He also saw a future when people traveled in "cars with sleeping accommodations and cuisines aboard, touring the country," in other words, RVs. He even predicted a kind of Internet: "As the citizen sits in his car, he may press a variety of buttons or turn an indicator and obtain any section he desires of the modern newspaper—the forests saved and millions of tons of waste paper eliminated."[20] Wright the visionary could be remarkably blinkered. The one thing he did not see coming was large-scale agriculture; instead, he assumed that farming would remain decentralized, with individual families growing much of their food on their one-acre lots.

It was inevitable that Wright, an architect, would give Broadacre City physical form. The opportunity arose in 1934, when Edgar J. Kaufmann, a Pittsburgh department-store magnate (and the future client of the famous Fallingwater house), commissioned him to build a large model of his futuristic city. Some historians maintain that Wright did not intend Broadacre City to be taken literally.[21] But the twelve-foot-square model, representing four square miles of vaguely midwestern countryside, with topographic detail, miniature buildings, and tiny cars, made a strong impact precisely because it was realistic.

Wright's vision is a curious blend of pragmatism and science fiction. A two-level highway (cars above, trucks below), with a high-speed monorail in the median, connects to a grid of country roads. The specially designed interchanges, which don't really function effectively, are compact versions of cloverleafs. Transportation is provided by strange-looking cars and "aerotors"— personal helicopters resembling flying saucers—likewise designed by Wright. Fourteen hundred families are accommodated in a variety of houses. Scattered apartment towers are halfway houses, for the "city-dweller as yet unlearned where ground is concerned."[22] The county administrative offices are in a skyscraper. Schools, motels, shopping malls, professional offices, factories, farms, vineyards, an aquarium, a large circular stadium, as well as a drive-in megachurch, are dispersed across the landscape. The random scattering must have appeared odd in the nineteen thirties, but it is immediately recognizable to anyone who has flown over Phoenix or San Jose.

The most radical aspect of Broadacre City is that it not only has no center but has no edge. The model Wright built is not a model of a complete city; it is a model of a small piece of what is obviously a much larger urban area that goes on forever, or at least until it runs up against a natural obstacle, such as a lake or a mountain. Previous ideal cities, even ones as radical as Le Corbusier's Radiant City, were based on a differentiation of city and surrounding countryside, but Wright described a world without country-side—and without cities.

Most architecture critics, who lived in cities, found his proposal outrageous and greeted Broadacre City with a mixture of derision and condescension. His admirers—embarrassed—turned a blind eye, treating the proposal as the harmless hobby of an aging genius. Several later biographers have referred to Broadacre City as a kind of private WPA program, implying that it was merely an excuse for Wright to keep his office busy during the Depression.[23] But work on Broadacre City continued long after the end of the Depression—and long after Wright's career revived. Over the next two and a half decades, he published two more versions of his urban thesis: *When Democracy Builds* and *The Living City*. And in the early nineteen forties, he designed a number of planned communities that incorporated Broadacre City ideas, albeit on a smaller scale: a workers' cooperative in Detroit, two housing developments near Kalamazoo, and an unbuilt project in Pittsfield, Massachusetts. His largest realized work, in Pleasantville, New York, was a subdivision of fifty-five three-quarter-acre lots on a hilly hundred acres.

What are we to make of Wright's planning ideas today? For Brendan Gill, Broadacre City remains a suburban nightmare, a "homogeneous non-city," and a "monstrously enlarged Oak Park," referring to the Chicago suburb where Wright once lived—although Broadacre City doesn't resemble Oak Park one bit.[24] Lewis Mumford considered *The Disappearing City* a "clear anticipation (romantically rationalized) of the contemporary exurban sprawl."[25] Others, such as Joel Garreau, the author of *Edge City,* admire Wright's "stunning accuracy."[26] But the importance of Broadacre City lies not merely in Wright's prescience. He turned away from traditional European urbanism, just as he rejected International Style architecture. Yet, whereas his dislike of European modernists such as Le Corbusier was in large part a matter of taste, his rejection of the conventional concentrated city was intellectually grounded. Although Wright was never as anti-urban as he pretended to be, he understood the direction in which the tide of urban history was running. He instinctively saw the economic advantages of decentralization for a country as large and dynamic as the United States. Perhaps he was rationalizing, as Mumford claimed. Wright underestimated the environmental effects of mass automobile use (though he did champion small cars, such as the Nash Rambler), but he was correct in assuming that an economy dominated by growth and technological change required a different, more flexible form of urbanism than the established cities of Europe. What he did not foresee was that the future Broadacre City would be shaped not by architects, as he intended, but by market forces.

Late in life, Wright published *The Natural House,* a small practical guide for prospective home builders. In a chapter titled "Where to Build," he advised on finding a building site. "The best thing is to go as far out as you can get. Avoid the suburbs—dormitory towns—by all means. Go way out into the country—what you regard as 'too far'—and when others follow, as they will (if procreation keeps up), move on."[27] Move on. The same sentiment draws people to Chester County today and is the driving force behind Arcadia's development. In that sense, Wright, no less than Raymond Unwin and Andrés Duany, must be considered New Daleville's spiritual godfather.

13

Design Matters

Most new residential communities don't have too little variety, they have too much.

Almost immediately after the New Daleville ordinance is approved, Jason Duckworth starts working on the so-called architectural guidelines. These are a sort of aesthetic building code that will govern the exterior appearance of the houses and ensure that New Daleville will turn out the way Arcadia—and the township—want. Most suburban builders are used to large lots, where houses are far apart—and far from the street—so the design of one house does not really affect its neighbors. At New Daleville, where lots will be small and houses very close together, the appearance of the houses will be critical. Hence the importance of the guidelines, which will dictate materials, colors, and details.

The idea of formally regulating the appearance of a place sounds artificial, but design controls were a part of garden suburbs from the beginning. Almost a century ago, Raymond Unwin taught that harmony is essential to a successful town plan and recognized that the planner would need to exercise some degree of control.

Where the site planner, to complete his street picture, requires even roof lines he should be able to suggest heights for eaves and ridges; where he desires to maintain definite colour schemes he should be able to suggest materials and treatment in accordance with these schemes. Where the position of his building requires a symmetrical, picturesque, or other special treatment to complete some effect aimed at, he should be able to suggest this treatment to the architect or builder who is responsible for the building on the plot.[1]

Unwin admitted that the "difficulties of such public control are undoubtedly very great," but he believed that "the evils which result from absolute lack of control are even greater."[2] Frederick Law Olmsted, Jr., echoed the sentiment: "Whatever may be done for the sake of color, enrichment or decoration, must be controlled by regard for the composition of these masses and open spaces or confusion will result."[3] At Palos Verdes, the detailed controls covered thirty-six pages, so that "every purchaser . . . may be sure when building his home there that his neighbor will have to build an equally attractive building."[4]

Architectural controls have taken different forms. Frank Lloyd Wright controlled the appearance of the Doheny development by designing all the houses himself. In Chestnut Hill, George Woodward commissioned architects who were sympathetic to his vision. At Palos Verdes, the "protective restrictions" that governed the appearance of the houses were enforced by a so-called Art Jury, composed of prominent local architects. Design review boards are a common feature of many modern planned communities, but when he was developing Seaside, Robert Davis thought a cumbersome review process would discourage future home buyers. Instead, he wanted the requirements to be clearly spelled out in advance so that they could be easily followed by local contractors. Andrés Duany and Elizabeth Plater-Zyberk admired old Florida coastal towns such as Key West and Apalachicola, whose architectural charm was the more or less accidental result of shared regional building traditions and a limited choice of building materials. But times have changed. Builders have access to a wide variety of natural and synthetic building materials, they can make roofs flat or sloped, and they can order windows of any shape or size. The result is that most new residential developments don't have too little variety, they have too much.

To limit choice, Duany and Plater-Zyberk created a set of rules that they called an architectural code, to distinguish it from the building code. On certain subjects, the Seaside code is unequivocal: the exact slope of principal roofs is fixed; white picket fences on the street side are mandatory; so are front porches. At the same time, the code leaves latitude for personal expression: it does not spell out the precise design of fences and porches, there is no limitation on architectural style, and houses can be painted any color. Although the height of buildings is strictly controlled, owners who want a view of the Gulf of Mexico can add a small lookout. Since their height, location, and design are unrestricted, the assortment of picturesque towers has become one of Seaside's characteristic architectural features.

Duany and Plater-Zyberk understood that Americans are attached to traditional architectural forms, such as pitched roofs and front porches. There is some scientific evidence for this cultural bias. Anton C. Nelessen, a New Jersey architect and urban designer, has developed a technique for assessing people's reactions to their surroundings. He calls it a Visual Preference Survey. Participants in the survey are shown a series of images and asked to rank them on a scale from most desirable to downright undesirable. The images include views of streets, houses, parks, and shopping areas. Ratings by different social groups in different parts of the country show a remarkable consistency.[5] Commercial strips, convenience stores surrounded by parking, and new residential developments on large lots in open land are always given negative ratings. Shopping streets and older residential areas with sidewalks and shade trees are always positively rated. "Adding landscaping to streets and parking lots, always results in a more positive rating," Nelessen observes.[6] He often uses paired images, and while people give low ratings to apartment complexes with large parking lots and bland buildings, they rate apartments with porches facing on-street parking highly. In short, Nelessen's survey shows that design matters.

Having an architectural firm write guidelines is expensive. At Fairstead, Arcadia's canceled project in Lancaster, the guidelines and the architectural plans cost $200,000. For New Daleville, a much smaller project, Jason Duckworth is saving money and writing the guidelines himself. "Londonderry Township has expressed a desire for New Daleville to have a distinctive local look," he tells me, "although I'm not sure that they, or we, know exactly what that will be." He photographs farmhouses in the area that could serve as models and includes the photographs in his first draft of the guidelines. Jason admits that he isn't sure how much "authentic" details are valued by home buyers, but he thinks a more traditional appearance will be an important marketing tool, since it will set New Daleville apart from competing developments.

When Tom Comitta reviews the draft, he points out that Jason's photographs are all from Montgomery County. He suggests that examples closer to New Daleville would be more appropriate. Jason agrees, but he is concerned about two things: the old houses in the hamlets near Londonderry are considerably smaller than what will be built at New Daleville, and they are rather plain. "We know that high-end Philadelphia buyers in the six-hundred-thousand-dollar range like houses that look like modified farmhouses, with interesting Colonial-style details," Jason says. "For less than half the price at New Daleville, we can't deliver the same thing, but I hope we can come up with something attractive and consistent."

The guidelines come up at the next monthly meeting of the planning commission. Tim Cassidy, who leads the discussion on the design of the houses, wants to know what input the township is going to have. "Basically, very little," answers Dave Della Porta. "We know what we're doing." He usually speaks his mind and sometimes sounds brusque. Cassidy is not satisfied with this answer. He wants the township to be involved. One of the other commissioners disagrees and says that, as a rule, they don't review the architectural design of residential developments. "But this is different," says Cassidy heatedly. "The architecture is an important part of the new concept that we've accepted. We need to know it's going to come out right." It's obvious that he feels strongly about this, and most of the commissioners seem ready to support him.

Della Porta isn't sure how to respond. Arcadia has convinced the township about the importance of design, so he can't really back away now. Seeing his colleague at a loss for words, Jason joins in the discussion. "It's the planning that is the most important issue," he tells the commission; "getting the architecture right is just icing on the cake." Cassidy strongly disagrees. He says that the houses will have to be top quality for the village idea to work. Then he reads from a draft of the guidelines. "It says here that houses should have classical proportions and show the architectural heritage of Pennsylvania. I agree with the general idea, but terms like *classic* and *heritage* are pretty fuzzy. So is the requirement that roofs should have steep slopes. The guidelines need to be much more specific." He quotes a section that lists synthetic stone, vinyl siding, and brick-textured concrete as acceptable "high-quality building materials." He says that he doesn't consider synthetic stone and brick-textured concrete to be high quality, and if vinyl siding is going to be used, it should be very limited. Cassidy hates vinyl.

Della Porta continues to argue that issues dealing with design should be left to the developers. "You don't want to get involved in architectural questions," he tells the commissioners. "You're not experts."

"Oh, but I am," Cassidy answers quickly. "And I care."

Unwittingly, Della Porta has blundered. Although Cassidy works in the Comitta office chiefly as a landscape architect, he is also a trained architect; in fact, he has a doctoral degree in the subject. So if there's an architecture expert in the room, it's Dr. Tim.

For the first time, there is a strong difference of opinion between Arcadia and the township. Cassidy refuses to budge, and he puts the Arcadia people on the defensive. The meeting ends without any clear resolution of the guidelines issue.

"I don't want the township to review every building permit application," Cassidy tells me afterward. "I just want the developers to build what the guidelines describe. I can't believe that Jason said that if the plan was good, mediocre architecture would be okay. In reality, anything they build probably *will* be mediocre. But if they shoot for mediocre, it will just be terrible!"

The vagueness in the guidelines that Cassidy objects to is not accidental. "Since I started to work on this, I've talked to a dozen different design professionals," Jason tells me. "I've asked them which are the three most important design rules, the real levers. Some say proportions, some windows, others say trim and moldings, or porches. Nobody agrees. There's no consensus on what is key." Is he having doubts about traditional neighborhood development? "I still think it's a better way to develop and build," he says. "But early on I wasn't sufficiently skeptical. The enthusiastic proponents of TND are often the people who are farthest from its implementation. I now have some questions about buildability and consumers' receptivity to the idea. TND doesn't satisfy every market, and it's not as easy to build as they say."

Jason has decided to leave the builders plenty of leeway. "The rules are vague because we don't really know how to specify good design," he says. "We asked people at the meeting to tell us what they wanted or didn't want, but we didn't get any clear answers. Tim doesn't like vinyl, but the kind of houses we're building will definitely have vinyl siding, so that's not very useful. Ultimately, it's the builders who understand what sells houses, and we don't want to tell them what to do."

Jason is in no rush. Nothing can happen until the engineers finish revising the preliminary drawings to reflect the new street layout. In the meantime, he receives a four-page letter from Cassidy with a long list of suggestions about how to tighten up the guidelines, and a heavily marked-up copy of the guidelines themselves. "No plastic fences!!!" reads one of the notes. Jason adjusts some of the language, but mostly he ignores Cassidy's suggestions. He's not worried about the planning commission at this point, since he's showing it the guidelines strictly as a courtesy. While there might be some modifications as a result of these informal discussions, he doesn't feel obliged to take Cassidy's advice.

A year earlier, when the subject of the design guidelines first came up, Comitta suggested that they should be formally adopted by the township. Jason resisted the idea and pointed out that the municipal planning code does not require such ratification.[7] Comitta agreed, but he now thinks he may have made a mistake. John Halsted, the township solicitor, says that, despite the county code, the new Londonderry ordinance is quite specific: the design guidelines must be adopted by a formal vote of the board of supervisors.

Jason has no choice but to agree. Halsted tells him that he doesn't want the township involved in overseeing detailed design issues—all that is needed is a simple "black and white" document that can be made a part of the ordinance. Since the guidelines will be a legal part of the ordinance, compliance will be determined by the supervisors, who will also have authority to approve—or disapprove—any future modifications. Thus, from Jason's point of view, the simpler the guidelines are, the better. He deletes a lot of the descriptive and photographic material and distills the twenty-three pages into eight. Comitta thinks the planning commission will be able to approve the guidelines at its next meeting.

Things do not go smoothly. "I thought that we had an under-standing, but we basically got a lot of push-back from the commission," Jason tells me. "Tim Cassidy feels strongly that the guidelines should be very specific, and the people in the audience rallied around him. One person even suggested that we should bring in samples of building materials, such as siding, and roofing shingles, for approval. That would be a disaster! Tim's position is that the township is taking a risk by giving us a density bonus, and in return it should have a greater say in design matters. Our argument is that the density bonus has nothing to do with the design of the houses. We're spending a lot on site improvements, granite curbs, trees, and so on. That should be enough."

The expected vote on the guidelines does not take place. Instead, Jason agrees to bring drawings of typical houses to the next meeting. He's getting boxed in, and he's not happy about it. "We've been going back and forth on these guidelines for several months," he says. "It's now sixteen months since the New Daleville ordinance was approved, and the project is more than a year behind schedule." Because of the delay, Arcadia has had to negotiate another extension with the landowner, pushing the option deadline from spring to the end of the year.

Jason sends a new draft of the guidelines to the township. At the last minute he includes illustrations of typical neotraditional houses, hoping to mollify the commission. Cassidy isn't impressed. "There are no dimensions, materials are not labeled, and there are no notes or specifications of any kind," he writes in response. "Many of the examples look just like typical suburban 'products.' What makes them better?" Jason deletes the illustrations.

When the engineers complete the preliminary plans, Arcadia formally resubmits the drawings, together with the new sewage planning module and the revised guidelines. At the planning commission meeting, Comitta presents a long list of comments and suggestions by the township engineer and various township consultants: the storm water consultants, the lighting consultants, and the traffic consultants. Della Porta, who is there with Jason, says he's willing to go along with all suggestions. "I think we're ninety-five percent in agreement. We're almost there." He wants to wrap things up.

But a new issue is introduced. Londonderry transportation consultants have recommended traffic lights and road widenings to handle the extra traffic generated by New Daleville and another new subdivision. Traffic lights are expensive—$100,000 each—and the township wants the developers to pay for a part of the so-called off-site improvements. This amounts to an impact fee, that is, a fee charged to a developer to cover the financial "impact" of his development. Local governments in many states use impact fees to pay for a variety of improvements to roads, parks and recreation facilities, public works, fire companies, and police forces. Sometimes, the improvements may be only remotely connected to the developer's project and amount to a development tax. In Pennsylvania, however, thanks to intense lobbying by developers and builders, the state planning code requires that, before an impact fee can be levied, the municipality has to carry out a study to assess actual costs. Since Londonderry has not done a traffic study, it's on shaky legal ground. Della Porta, who knows this, says he will contribute to traffic improvements, but he cagily does not specify an amount. He leaves the delicate subject of the design guidelines until last. He says that, since everyone is in basic agreement—which is not quite true—there should be no problem ironing out the details. It looks as if he hopes to bluff his way through.

From the ensuing discussion, it's obvious the commissioners are divided. One group is ready to approve the project tonight, while the other, led by Cassidy, wants all the details—including the guidelines—resolved first. At one point, it looks as if the motion will be shelved. Then Richard Henryson, the chairman of the commission, weighs in. "We have to have a little faith," he says. He suggests that a vote be taken on the preliminary plans, contingent on the various details being worked out. This is reasonable, and Cassidy doesn't object. Everyone falls in line, and the vote in support of New Daleville is unanimous.

When I later speak to Joe Duckworth, he is relieved. The project is running behind schedule, but that's normal. They are still not out of the woods. State approval of the drip irrigation scheme is a major hurdle. He thinks Jeff Miller has done a great job preparing the sewage module, but the Department of Environmental Protection is unpredictable. Still, it looks as if his first impression of Londonderry was correct — with the exception of Cassidy, the township has been cooperative, and the chances for New Daleville look good. Duckworth is also very pleased with his son. "He can really handle things on his own now," he says. "My only worry is that he sends me e-mails at two forty-five in the morning. He works too hard." Duckworth speaks to Jason's wife, Angela. "Jason has worked equally hard in all three of the jobs he has had since graduating," she says. "But this is the one that he likes the best."

"I do like it," says Jason when I tell him. "It's not just working with my dad, which I enjoy, but the job itself. It's a business, but it's also a creative endeavor. There are so many sides: political, aesthetic, environmental, even personal." Is there anything about it he doesn't like? "I was raised to be a good kid, to do the right thing and try and please people. What I don't like is being the bad guy. Sometimes I have to make people unhappy to get the result we need." I ask him what has surprised him most about being a developer. "The most valuable skills in my previous jobs were analytical. Real estate development values pragmatism and the ability to quickly solve problems in a commonsense way. I'm not sure what the value of an Ivy League degree is in all this. In real estate you have to get the big things right, like acquiring the land and getting permits, but to actually cross the finish line you have to deal with a hundred small details."

Locked In

A recent Virginia study found that homes that were part of a community association "typically sold for 5 percent, or $14,000, more than a similar home nearby not governed by a homeowner association."

Municipal zoning was not widespread until after the nineteen twenties, so the builders of the early garden suburbs needed a way to guarantee the implementation of their plans—and to preserve property values. They adopted two legal mechanisms: the deed restriction and the home association. Deed restrictions, also called restrictive covenants, were clauses attached to private lot deeds that were binding on the original buyers and on all subsequent owners. They controlled what could be built on the lot, in terms of size, use, and quality.* The home association, to which all owners belonged, enforced the deed restrictions and also owned and maintained common areas in the development, such as parks and other public spaces.

Although all the classic garden suburbs—Riverside, Forest Hills Gardens, Mariemont, and Palos Verdes Estates—were gov-

*Deed restrictions are implicitly exclusionary, since not everyone can afford to meet their demands. Some of the early restrictions were explicitly exclusionary on the basis of race. In 1948 the Supreme Court ruled racial deed restrictions unenforceable.

erned by home associations, for a long time such organizations remained upper-middle-class rarities.* Today, 30 million Americans belong to home associations, called homeowner or community associations, and it has been estimated that, in the largest metropolitan areas, more than half of all new homes are governed by an association.[1] Modern community associations are the products of a single event. In 1978 California taxpayers approved a statewide amendment—Proposition 13—that reduced local property taxes and limited future increases. The initiative spread nationally, and within five years over half of the states had adopted similar legislation. "Twenty-five years ago, we'd build parks, green spaces, and retention ponds, and when the project was finished we'd deed all these public spaces to the municipality," says Joe Duckworth. "The taxpayer revolt changed all that. Municipal governments everywhere refused to take responsibility for maintaining public spaces in new developments. They basically told us, 'Take care of it yourselves.'" Henceforth, public amenities would remain the responsibility—and the property—of homeowners. According to Duckworth, before Proposition 13, community associations were rare in Pennsylvania. "Now, *all* our projects include community associations. The local municipal governments insist on it."

A community association is a private, not-for-profit corporation, founded by the developer with the approval of local government, to which all the homeowners in a development automatically belong. The association, which is run by a board elected by the individual owners, collects monthly dues from each member. The purpose of these dues is to pay for the upkeep and repair of common property, such as swimming pools, clubhouses, tennis courts, golf courses, parks, and playgrounds. Some associations also own streets, sidewalks, and parking lots.

---

*George Woodward did not use deed restrictions in Chestnut Hill because most of the houses he built were rented rather than sold.

Although all municipalities have zoning today, deed restrictions have not disappeared. These restrictions go much further than most municipal ordinances. For example, New Daleville home-owners will not be allowed to erect satellite dishes on their front lawns, or park recreational vehicles, trailers, or boats where they are visible from the street, or store trash, lumber, old cars, or scrap in their backyards. Nor will they be allowed to build above-ground swimming pools or any sorts of accessory buildings, such as toolsheds or playhouses. They will need association approval to plant shrubbery more than three feet high and to put up flags, banners, or signs. And despite New Daleville being in the country, they won't be able to keep livestock.[2] Such rules are not unusual. "I actually haven't read the New Daleville agreement," Duckworth admits. "Most of it is pretty standard. It's what we and our lawyers, who draft these things for a living, have found works. There are some buyers who don't want any restrictions, but most people accept reasonable limits on what they consider objectionable behavior. They don't necessarily read the fine print when they buy the house, but believe me, they do when their neighbor does something obnoxious."

"Once we finish the project, people can decide to change the restrictions," he says, "but they never do. I don't know of any case where they've changed anything fundamental." That is hardly surprising. At New Daleville, it will require the support of at least two-thirds of the owners to pass an amendment and a unanimous vote to terminate the agreement. The agreement will run for thirty years, which is typical, at which point it will automatically be extended for ten more years, unless terminated by a vote of not less than 80 percent of the owners. In other words, the New Daleville association will be around for a long time.

In the popular imagination, community associations are often linked to gated communities. This perception has lead to a series of critical books with alarmist titles such as *Privatopia, Behind the Gates,* and *Fortress America.* But not all communities governed by an association are gated. According to the U.S. Census Bureau, although some 10 million American households belonged to community associations in 2001, fewer than 4 million lived in gated communities.[3] This represents about 4 percent of the total population, a small fortress indeed. Like most developers, Duckworth is pragmatic about gated communities. "I've built only one gated community in the Philadelphia area," he says. "It was an upscale community with a hundred and twenty homes, which would each probably sell for about half a million dollars today. We paid for the gates, built a guardhouse, hired guards. I live nearby, and I noticed that shortly after we handed over the project, the association fired the guards and left the gates permanently open." He believes that, at least in the Northeast, gates are more about projecting an image of exclusivity than about maintaining security. By contrast, he says that if you are building a planned community in South Florida, or in many places in California, buyers absolutely demand walls and gates. "I think gates are a bit like doormen in urban apartment buildings. Are they there for the image, the convenience, or for security?"

There will not be any gates at New Daleville. Access to the neighborhood will be public, since all the streets, although built by the developers, will be deeded to the township when the project is finished. "We insist on that," says Duckworth. "After all, the owners pay property taxes, and they shouldn't be burdened with policing, snow plowing, and repaving the streets." The association will be responsible for maintaining the rear lanes, the green courts, the walking trails, the play lot, and the storm water retention basins, and for the landscaping in the common areas.

Community associations are governed by the homeowner agreement. The sixty-five-page New Daleville agreement is the responsibility of Simi Baer Kaplin, whom I meet for lunch in a suburban mall. She graduated from the University of Pennsylvania Law School four years ago and works at Kaplin Stewart, her father's law firm. I ask her how she decides what to include in an agreement. In Pennsylvania, community associations follow state regulations, she says, so much of the text is standard: how the developer establishes the association, how the association is run, the rights of the various parties, the procedures for compliance and default, and so on. However, what she calls the "lifestyle restrictions" vary from project to project. They depend on the experience of the developer, what other developers in the area are doing, and the kinds of home buyers the development is aimed at. According to Kaplin, the restrictions are a combination of what buyers expect, what makes them comfortable, and what doesn't discourage them from buying.

"I asked Dave Della Porta how strict he wanted the rules for New Daleville to be," she tells me. "He didn't seem much concerned and left it up to me. It's housing for families, so that influences the restrictions. In general, I try to be reasonable. For example, I think that restrictions on clotheslines not being visible from the street are okay here, but I wouldn't write that kind of rule for a modestly priced town-house development where not all the units have dryers." The restrictions also reflect her own biases. "I tend not to write complicated rules about hiding trash cans," she says. She agrees with Duckworth that most buyers probably don't read the agreement at the time of sale. "After all, it's more than sixty pages. I belong to a homeowner association, and I've read all the rules because I'm a lawyer. But my husband, who's a podiatrist, doesn't care."

Kaplin points out that community associations can be activist. "It's not uncommon for associations to sue developers who they believe have cut corners in completing a project." Associations can also be strict with their own membership. The board of directors has the right to levy fees on the owners, and to assess fines and late charges, although unlike co-op boards, they do not control who becomes a member. Because associations have such wide discretionary powers, they are sometimes darkly referred to as "private governments."[4] They do not seem so different from social clubs, religious denominations, and other groups that also have special rules about private behavior. But there is no doubt that community associations represent a privatization of traditionally municipal responsibilities, such as street cleaning, trash removal, and park maintenance.

Such privatization is not confined to suburban planned communities. Business improvement districts, popular in many downtowns, likewise do their own cleaning, street maintenance, and policing. Some large city parks are maintained by privately supported conservancies. The privatization of public space has raised concerns among academic critics, but "public space and publicly-owned spaces are not the same thing," writes Richard Briffault, a law professor at Columbia. "Some of the most successful public spaces—successful in the sense that they contribute to a flourishing public life—are privately owned, such as the English pub, the French sidewalk cafe, the Viennese coffee house, the German beer garden, or closer to home, Rockefeller Center, South Street Seaport, the Quincy Market Place or the Columbia University campus."[5] The village green that Bob Heuser has planned for New Daleville will be similarly publicly enjoyed, although it will be owned by the homeowner association.

Jason Duckworth is responsible for squaring the New Daleville homeowner agreement with the township. "The agreement is actually an entirely private document," he tells me. "It's strictly between us and the home buyer; the township is not party to it. But we're asking them to review it, since we need to make sure that our regulations are in line with their ordinances. For example, if Londonderry Township didn't allow pink flamingos on front lawns, we couldn't permit them in New Daleville." Bob Harsch, the township engineer; John Halsted, the solicitor; and Tom Comitta all comment on the agreement. Harsch's main concern is the association's ability to manage the sewage treatment system. If there were a major failure, he says, the neighbors' water supply might be affected, and the township could be held responsible for repairing the system. Jason assures him that the sewage treatment system will be owned and operated by a public utility. In an arrangement that has become increasingly common, the utility will buy the system for a nominal sum from Arcadia and collect fees for operation and maintenance from the community association. Halsted wants to make sure that the association will be properly indemnified. Comitta's concern is that the agreement be compatible with the traditional neighborhood development ordinance. For example, he wants it made clear to homeowners that, since the plan of New Daleville is subject to a municipal ordinance, the association cannot make any significant changes without the approval of the township. He also has some concerns about the landscaping restrictions, which limit the height of what people may plant in their gardens without the approval of the architectural review committee. Kaplin thinks he is worrying needlessly. "Do I know the difference between twenty-four-inch and thirty-six-inch shrubs? No. Do I care?"

The homeowner agreement refers to a "Handbook for Builders and Contractors." This document, which will govern the actual design and construction of the houses, contains the material that Jason removed from the old design guidelines. In reviewing the agreement with Kaplin, Comitta suggests that a clause be added to the effect that any changes to the handbook would be "subject to the consent of the township." Although Kaplin sometimes flippantly refers to New Daleville as "Ye Olde Neighborhood," she has worked with Arcadia on several projects and understands that it is important for developers to have flexibility as construction progresses. "The design rules are determined by a market analysis and may require revision based on sales figures and reactions by the public," she writes to Comitta. "My clients cannot allow the Township to have control over what will be a business decision." She is unyielding on this point.

Community associations give developers a high degree of control while houses are being marketed and sold. At New Daleville, Arcadia will establish the association and appoint the board of directors before any houses go on sale. Since there are no homeowners, the board will consist of Christy Flynn, Jason's associate, as president, and Susan Callaghan, Arcadia's financial manager, as secretary and treasurer. As houses are sold, power will gradually shift to the homeowners, although majority control will remain with the developer. "Our first concern with the restrictions is to regulate the appearance of the development during the marketing period," says Duckworth. "If a new owner breaks one of the rules, like parking an RV on their lot, or putting up a prefab storage shed, we'll have a sales associate call them and point out very politely that they're out of line with the agreement that they just signed. I can only remember one or two instances where we've ever had to take further action." According to Pennsylvania law, once three-quarters of the houses are occupied, the developer must cede full control of the association to the homeowners. "After we hand over management of the association to the owners, they can do what they want," Duckworth says. "What they often do is to fire the landscaper and get someone cheaper, and less good. But that's their business."

Community associations give homeowners a direct sense of control that is often—sadly—lacking in urban and suburban neighborhoods. The association rules are restrictive, but they apply equally to everyone. Unlike municipal zoning ordinances, which can be altered—or circumvented—deed restrictions are difficult to change. This inflexibility may prove to be a liability in the long run, since densification or future subdivision of large lots will probably be impossible, but in the short run it gives homeowners confidence that property values will be protected. Which explains why associations are so popular. According to a 1999 Gallup survey of community association members, fully three-quarters felt that their association rules were either "very" or "extremely" appropriate for the community. The survey also questioned homeowners who did not live in a community association. A scant 7 percent said that they wouldn't join an association because they objected to "too many" rules and regulations.[6] Moreover, there is evidence that buyers place a premium on private government. A recent Virginia study found that homes that were part of a community association "typically sold for 5 percent, or $14,000, more than a similar home nearby not governed by a homeowner association."[7]

Although Jason Duckworth and Dave Della Porta have had many meetings with the planning commission, the preliminary plans for New Daleville have still not been approved by the Londonderry supervisors. In preparation for a vote, the township consultants and the developers meet in the township solicitor's office to hammer out the final details. Kaplin's father accompanies the Arcadia team. "Marc Kaplin has a reputation as an exceptionally tough and aggressive lawyer," says Joe Duckworth, who is not at the meeting. "Having him there lets the township know that we're serious."

The meeting lasts three hours. "We went through every item," Comitta tells me. "If it was legal, the lawyers discussed it; if it was technical, the engineers came up with an answer. At the end it seemed like almost everything was resolved." Even the contentious design guidelines. It is agreed that the guidelines will not include drawings of typical houses—Jason gets his flexibility after all. But he does not get off scot-free. Arcadia agrees to hire an architectural firm to oversee the application of the guidelines. The township solicitor insists that, before the final plan of New Daleville is approved, the guidelines will be formally adopted by the township as an ordinance, and that all discussions about the guidelines must include Tim Cassidy.

During the meeting, the developers are surprised to discover that Londonderry requires them to pay a fee of $2,000 per house in lieu of providing recreation space. "It's amazing that they never told us about it," says Jason, "but since this ordinance was put in place just before we submitted our plans, we are liable." The township agrees that the cost of grading the neighborhood park and putting in walking trails and play lots will be counted as a credit, which will substantially offset the recreation space fee. The issue of the cost of off-site road improvements is still up in the air. Della Porta makes a cash offer of $65,000 on the spot, but the township says it prefers to wait until the traffic consultants have completed their study.

The supervisors' meeting to give preliminary approval to New Daleville is chaired by Fred Muller, since Howard Benner has stepped down after serving his four-year term. The new supervisor, Martha Detering, a trim woman in her early fifties, was elected last fall. "Marty lives directly across the road from me, the horse farm with a creek running through it and a riding ring," says Cassidy. "She's a retired schoolteacher, mother of two, and a hardcore equestrian enthusiast. Her husband, Hank, is a retired Marine pilot and principal of the local high school. They're the best neighbors anyone could ever have, and I'm delighted that she has taken the thankless responsibility of being a township supervisor."

Della Porta is the only one of the New Daleville team in attendance. There's a handful of people in the audience. A young man in a wool jacket comes in. "We just got a quorum," says Muller. "Let's start." Della Porta gets up and says that, except for the design guidelines, which will be voted on later, all the conditions for preliminary approval have been agreed upon, and there is no need to go over them again. "What about this blank space here?" asks Muller. He is referring to the section that describes New Daleville's contribution to the off-site road improvements—the dollar amount has been left open. "Do you agree with the consultant's report that suggests one thousand to fifteen hundred dollars for each peak-hour trip generated by New Daleville?" Detering asks Della Porta. According to most traffic engineers, a single-family suburban house generates one car trip per day during peak hours. That means New Daleville could have to pay between $140,000 and $210,000.

"A thousand dollars sounds like the right number," Della Porta says warily. "We thought of one thousand as the lower figure," says Detering briskly. Della Porta answers that he isn't aware of other projects in the area where improvement costs have been higher than a thousand dollars per trip. Now they're horse trading. It's clear that the two supervisors (Clair Burkhart, the third supervisor, is absent) have not discussed this between themselves, but Della Porta presses them for a decision. "Since you're going to make a traffic study, could we agree that we will pay a thousand dollars, and if the costs are higher, then we'll pay the extra cost, up to a ceiling of fifteen hundred?" he asks. "I'm happy with that," says Detering. Then Muller makes a formal motion to give preliminary approval to the New Daleville plan. Detering seconds, they vote, and it's done. The whole matter takes less than fifteen minutes.

For developers, receiving preliminary approval for a project is the critical step in an entitlement process. "Once preliminary approval is given, the township can't back out or demand any substantive changes," explains Della Porta after the meeting. "As long as we follow the conditions, they pretty much have to approve the final plan. They're locked in."

Even before the negotiations about preliminary approval were concluded, Arcadia's engineers had started working on the final plans. Della Porta hopes to submit them to the township in a couple of weeks. The sewage module will be ready for the next meeting of the supervisors, who will then forward it to the Department of Environmental Protection. The state agency is supposed to respond within sixty days, but according to Jeff Miller, the environmental engineer, "never in history has DEP approved a drip irriga-

tion project that quickly; they typically extend the review period."
He guesses that it may take as long as six months for the sewage
module to be approved.

More delays mean higher costs. On the one hand, with the
modifications to the drip irrigation system, the unexpected off-site
road improvements, and the added public amenities, such as the
recreation area, the total cost of New Daleville has increased significantly. Della Porta's latest pro forma, or financial calculation,
shows that the total cost of an improved lot is now considerably
more than the original estimate. "This is moving from a high
number to an even higher number," says Jason. On the other
hand, during the intervening two years, house prices have also
risen. "We've been studying new communities in the area, and we
believe that our house prices will start in the mid–two hundred
thousands, and go up from there."

The increase in house prices is driven by two factors. First, a
strong economy has increased the demand for home ownership,
and low interest rates have raised the prices that buyers are prepared to pay. Second, the price of land is going up even faster. The
obstructionism of many townships, coupled with stricter environmental controls and a lack of infrastructure, has produced a
scarcity of permitted land in Chester County. As a result, builders
are prepared to pay more for lots. "When I entered the business in
1976, the lot represented only fifteen percent of the selling price of
a house," Duckworth says. "It was about twenty-seven percent
two years ago, now it's up to thirty percent. In some parts of the
country, it's as high as fifty percent." In the case of New Daleville,
that means that if builders expect to sell houses for $250,000, they
will be prepared to pay as much as $75,000 for a lot. This will cover
the increase in cost, but Della Porta's pro forma indicates that
with all the extra expenses, New Daleville's rate of return has fallen
from 17 percent to less than 14. "The project still works, although
the profit margin is tighter than we would have liked," he says.

15

House and Home

Three-quarters of Americans live in houses. Contrary to popular opinion, this is not a reflection of suburbanization, since most city dwellers live in houses, too.

New Daleville will consist entirely of houses. Four out of five of all new housing units currently built in the United States are single-family houses.[1] This statistic has less to do with the nature of the home-building industry, or the suburban location of new housing, than with buyers' preferences, that is, What People Want.

I've lived in a house during most of my life. In Scotland, where I was born, my mother rented a room in a private village home (my father lived in army barracks). Later, we shared a large London house with a group of Polish émigrés. Eventually, my parents bought their own terrace house in a Surrey suburb. Growing up in Canada, I lived in a small-town bungalow. Except for a ten-year hiatus, when we lived in a flat in downtown Montreal, my wife and I have always lived in a house—first rented, then built, and last bought.

Many things—government policies, tax structures, financing methods, home ownership patterns, and availability of land—account for how people choose to live, but the most important factor is culture. To understand why we live in houses, it is nec-

141

essary to go back several hundred years. Rural people have always lived in houses, but the typical medieval town dwelling, which combined living space and workplace, was occupied by a mixture of extended families, servants, and employees. This changed in seventeenth-century Holland. The Netherlands was Europe's first republic, and the world's first middle-class nation. Prosperity allowed extensive home ownership, republicanism discouraged the widespread use of servants, a love of children promoted the nuclear family, and Calvinism encouraged thrift and other domestic virtues. These circumstances, coupled with a particular affection for the private family home, brought about a cultural revolution.[2] People began to live and work in separate places; children now grew up with their parents (rather than being apprenticed to strangers, as before); and the home, securely under the control of the now "housewife," was restricted to the immediate family. This intimate family haven was always a house. Seventeenth-century Dutch cities and towns were composed almost entirely of houses built in rows, side by side, wide or narrow depending on the wealth of the owner.

The idea of urban houses spread to the British Isles thanks to England's strong commercial and cultural links with the Netherlands. In the eighteenth and nineteenth centuries, Britain adopted the row house in many guises, as Georgian crescent, middle-class terrace, and workingmen's row. It has been estimated that, by the beginning of the twentieth century, nine out of ten dwellings in England and Wales were row houses.[3] Today, the vast majority of the British population—four households out of five—live in houses.

"A man's home is his castle" is a basic concept in Anglo-Saxon common law. For example, the courts have upheld rights of privacy which allow conduct in the home that might be prohibited elsewhere. According to D. Benjamin Barros of Widener University, the so-called castle doctrine gives special legal protection to the security of the home. "First, the law privileges certain acts of self-help made in defense of the home that would in another context be criminal or tortious. Second, the law imposes criminal sanctions upon individuals who invade a home, and these sanctions are significantly greater than those imposed for invasions of other types of property."[4] The primacy of the home in law originated in Britain but migrated to the United States, where the Third and Fourth Amendments to the Constitution prevent the government from unilaterally quartering soldiers in private houses and provide special protections to houses in search and seizure law.

Modern Holland has remained a predominantly house-occupying country, as has Belgium, which was originally a part of the United Provinces of the Netherlands. In Norway, which historically had close trade ties with Britain, a large majority also lives in houses.[5] The United States (as did Ireland, Canada, and Australia) inherited the Anglo-Dutch house tradition, and three-quarters of Americans now live in houses.[6] Contrary to popular opinion, this is not a reflection of suburbanization, since most city dwellers live in houses, too.[7]

It's one thing to say that people prefer to live in a house, but what *kind* of house? Basically, there are three choices: a freestanding house, a house sharing common walls with its neighbors, and a house that is oriented to an inner court. The last is an ancient model. The Roman dwelling was the classic courtyard house. Generally one story high, it covered the entire lot. Depending on its size, it had one or several open-air courtyards. The entrance court, called the *atrium,* led to the *peristylum,* a large court surrounded by an arcade. The latter, unlike the paved *atrium,* often contained a garden. The courtyard house, small or large, was the dwelling of choice; only the poorest Romans lived in *insulae,* or multistory tenements.

The courtyard house is common throughout the Mediterranean. In Spain, the court is called a *patio,* a domestic feature that was transported to Latin America. Although the courtyard house is efficient in terms of use of land—a courtyard house has no setbacks, usually not even a front yard—the court provides absolute privacy to the occupants. Courtyard houses are also found in Asia. Until the Communist Revolution, which introduced Soviet-style apartment blocks, Chinese cities consisted almost entirely of courtyard houses, which date back to the Han dynasty, about 200 B.C.[8]

Courtyard houses are limited to one or two stories; otherwise the courts become too dark. If land is at a premium, as it was in seventeenth-century Amsterdam, three- or four-story row houses are a higher-density alternative. The occupants of a row house must make concessions, however. The interior layouts will be constrained by the long, narrow shape and the limited number of windows. The backyards can be overlooked by neighboring houses. Privacy—visual and acoustic—is reduced. Semidetached houses, sometimes called twins, which are attached to only one neighbor, mitigate some of these defects, since they have more windows and wider yards. In Britain, once a nation of row houses, today approximately one-third of all houses are semidetached.[9]

In America, as in Britain, row houses were a common feature of nineteenth-century industrial cities. Today, about half of all houses in metropolitan Philadelphia are still row houses. However, prosperity has given Americans other options, and the row house, or town home, has fallen out of favor in postindustrial cities. In metropolitan Houston and Los Angeles, for example, only about one house in ten is a row house.[10]

Ninety percent of single-family houses in the United States are detached (in cities, the proportion is only slightly lower). The advantages of detached houses are many: light and air from all sides; greater acoustic and visual privacy; less danger of fire from neighboring buildings; and being able to pass from the front yard to the backyard without going through the interior. Even if the lot is only slightly wider than the house, the difference in terms of privacy is significant. Typically, buyers will pay a 10 to 15 percent premium for a detached house over a row house, even if the floor areas are identical.

Americans are hardly alone in favoring freestanding houses; indeed, they could be said to be typical. The first African town I visited, in 1982, was Makurdi, in south-central Nigeria. This sprawling city on the Benue River had about 100,000 inhabitants. I spent a lot of time walking the unpaved streets, blocked with uncollected garbage, which was the reason I was there—our team of Canadian consultants was advising the government on municipal sanitation. The residential neighborhoods consisted of low, freestanding houses on large lots. Although the houses were surrounded not by lawns and flower gardens but by vegetable plots and chicken coops, the meandering streets were shaded by huge trees. In an odd way it reminded me of a garden suburb. Since Makurdi was laid out by British colonial planners in the nineteen twenties, the resemblance was not accidental.

Cities composed largely of houses are common in sub-Saharan Africa, where, as in America, land is plentiful and population density is low. That is not the case in South Asia. But even there, given a choice, people have opted for houses.* In New Delhi and Madras, well-to-do Indians occupy neighborhoods of freestanding villas; the poor live in slums—but in row houses. The same pattern is visible in Manila and Bangkok. In 1986 I visited a recently built housing development in Hong Kong's New Territories. The developer proudly showed me around Fairview Park, five thousand small semidetached "garden houses." Hong Kong itself is a city of apartments, but Fairview Park had many of the attributes of an Anglo-American garden suburb: landscaped streets lined by individual homes with garages and private gardens.

During the same trip I visited mainland China. In Shanghai, I

---

*Cities in North Africa and the Middle East, like those of continental Europe, tend to be made up of multifamily buildings rather than individual houses.

was taken around extremely bleak apartment blocks belonging to the university. I asked my host, a professor, if he could show me any privately built housing. We drove to a residential neighborhood at the edge of the city. The owners were prosperous farmers who had invested their earnings in their homes—all freestanding houses. The spacious interiors were much larger than the two-room apartment that my host shared with his family. The economic revolution that would sweep the country was only just beginning. I doubt that the homes of university professors have changed much today, but for the growing entrepreneurial class, housing choices have expanded dramatically and now include American-style suburbs with single-family houses.

Even in countries such as France, Germany, and Russia, where many people still live in apartments, the number of single-family houses is growing.* "Polls consistently confirm that most Europeans, like most Americans, and indeed most people worldwide, would prefer to live in single-family houses on their own piece of land rather than in apartment buildings," writes the University of Illinois professor Robert Bruegmann.[11] It is the global nature of this desire, as much as the Anglo-Dutch tradition, that explains the popularity of single-family housing in the United States. Fast food, Hollywood movies, and American professional sports are a matter of taste, but most immigrants don't have to be sold on the idea of the individual house. It's a universal preference.

---

*The housing stock in France and Denmark is equally divided between houses and apartments. In Sweden and Finland, where government policies have discouraged building houses, fewer than half of families live in houses. In Italy, Spain, and Germany, the figure is even lower, although the number of houses has been growing.

Generic Traditional

*Theming provides a coherent and instantly recognizable set
of visual cues, to the home builders as the development is
being created, and later to the people who live there.*

The Chester County planned community of Weatherstone
will have more than 270 houses when it's complete, although
now there are only about 30. All the familiar neotraditional features are here: small lots, houses close to the street, garages in the
rear, porches, sidewalks. Everything is brand-new—the porches,
the sidewalks, the street signs, even the grass looks new. There are
some attractive details, such as the basement walls, which are
brick rather than concrete and have molded brick caps. But every
single house has exactly the same brick basement, the same cap.
There are streets in Chestnut Hill where all the houses were built
by George Woodward, but he used different architects and different builders, and made no attempt to standardize the house
designs. Here, by contrast, the development, design, and construction are by the same company, so even the attempts at variety have
a sense of sameness.

I'm waiting for Jason Duckworth and Tim Cassidy. They have
spent the day visiting neotraditional developments to look at
what local builders are doing. "The goal is to reduce anxiety and
build confidence with Tim," Jason told me earlier. Two of the four

projects that they are visiting are in Bucks County; the others are in New Jersey and, last, here in Chester County.

Jason's station wagon rolls up. He and Cassidy are accompanied by Christy Flynn, the Arcadia intern who has graduated and is now working with Jason full-time. "We've seen a bunch of projects today," says Cassidy. "It's a bit like looking at suits on a rack." I ask him his chief impression. "Inconsistency," he says. "One builder will use nice shutters but really cheap trim. Another will use a solid-looking bay window but flimsy railings." We examine one of the houses. The sturdy windowsills are made out of rowlock headers, bricks laid side by side in a traditional pattern. But the pediments over the windows are flimsy plastic profiles that are obviously screwed to the brick wall. "Why did they have to do that?" says Cassidy. "It would have been better if they had just left them off." He walks around the development, pointing out good and bad features, obviously enjoying himself. He can't help pontificating, it's the academic in him.

Jason says that an interesting aspect of the tour has been seeing how differently various builders did the same thing. One project they visited had concrete porches, which were cheaper than wood. Another builder used wooden porches but substituted vinyl for wood railings. Vinyl is cheaper than wood, and is a low-maintenance material, which appeals to buyers. "Plastic railings look okay from a distance," he says, "but when you get close and touch them, they just feel so bad." Flynn points to the stairs of a porch. "A lot of builders use these synthetic wood treads, I think they're made out of recycled plastic or something. They actually look okay, especially when they're combined with wooden risers, as they are here. They certainly fooled me."

One subject on which Cassidy and Jason don't agree is what Jason calls the four-sided versus the one-sided approach. Most builders use one or the other. The houses at Weatherstone are four-sided, that is, the builders have used the same materials and details on all four sides of the house. This appeals to Cassidy's architectural sense of consistency. In another of the developments they visited, the builder used more expensive details and materials on the front of the house and cheaper alternatives on the sides and back: vinyl siding instead of brick, plain details, basic windows. Jason agrees that the contrast between the front and the rest of the house can be jarring, but he thinks one-sided houses greatly improve the appearance of the street.

Cassidy and Jason seem to be getting along. "We want to keep you involved," Jason says. Cassidy appreciates that the developers really want to do something better than average; at the same time, seeing actual houses brings the conversation down to a concrete level. "It's the first houses you build that are important," he observes. "They will set the tone. If you can get them right, the other builders will follow. But if they're mediocre, it will only get worse," he adds.

Jason is generally pleased with Cassidy's reaction to the tour. "There were two subjects that I thought we would talk about," he tells me the next day. "The first was to identify good building techniques for New Daleville in terms of materials, details, and proportions. I think that's gone really well. Tim has already made some useful suggestions about the architectural guidelines. He doesn't like artificial stone, for example, and we're going to require that artificial stone can only be used under specific conditions. Tim and I agree on a lot of things."

The second subject was architectural style. "When we started New Daleville, which is my first project, I felt very strongly about identifying a regional style that would be part of the architectural code," he says. "Now I'm not so confident that we really know how to do this. Part of the problem is that there is such a wide range of styles in southern Chester County. There are ornate Victorian houses, classical Colonials, and plain farmhouses. Some of the towns have lots of brick, but some don't. Fortunately, during this tour we got so wrapped up looking at building techniques that the subject of style didn't come up."

There is no general consensus about what role style should play in the design of neotraditional neighborhoods. Andrés Duany has tended to dismiss the importance of style. When he and Elizabeth Plater-Zyberk published their impassioned critique of contemporary planning, *Suburban Nation,* they were unequivocal: "Traditional neighborhood design has little or nothing to do with the issue of architectural style."[1] Yet a big part of Seaside's appeal has to do with its appearance. "Of course Seaside is fundamentally an exercise in nostalgia," wrote Kurt Andersen in a sympathetic *Time* essay. "It manages to conjure the good old days impeccably, solidly, jauntily, even profoundly."[2] Developers understand the appeal of old-fashioned architecture. When builders up and down the Gulf coast copy Seaside, they make sure that their houses have picturesque porches, pastel colors, and tin roofs.

The Charter of the New Urbanism, which is the mission statement of the neotraditional neighborhoods movement, states that new developments "should respect historical patterns, precedents, and boundaries," and that architectural design "should grow from local climate, topography, history, and building practice." That sounds sensible; however, as Jason has discovered, there are problems with this approach. The first generation of neotraditional developments was mainly in the South, which has a rich architectural heritage. But not all regions have an identifiable architecture. As the United States was settled, immigrants brought their architecture with them, and in a country where people were so mobile, "historical patterns" and local "building practices" became hopelessly muddled. Even in older places that are associated with specific building traditions, such as Cape Cod, there is a wide range of domestic styles: Victorian, Queen Anne, Richardsonian Romanesque, and Levittown-era modern, to name but a few. As for the Cape Cod cottage, it has long since outgrown its regional roots and become a national style, as much at home in Michigan as in Massachusetts.[3]

Neotraditional developments fall into two camps with respect to style. Some adopt a more or less uniform style that evokes a particular time and place. This trend began at Seaside, and the person responsible, as much as anyone, was Melanie Taylor. She was born and raised in Miami, and before attending Yale architecture school she studied urban design at the University of Miami. There, one of her teachers was Andrés Duany, who encouraged her to become an architect. In 1982, when Seaside was beginning, she had just opened an office in New Haven with her husband, Robert Orr. When Duany heard that Orr & Taylor was looking for clients, he suggested they talk to Robert Davis. "We drove down to Seaside," Taylor recalls. "There was hardly anything there: a sales office, Davis's own house, and several cracker shacks that had been moved onto the site to fill up the one and only street, Tupelo Street." Davis offered the young architects a commission, to design houses for six lots at the very edge of the site. The land faced a neighboring subdivision, and Davis figured that he would be able to sell the lots only if they came with interesting-looking houses. Taylor and Orr suggested something different: a cluster of fourteen small cottages centered on a common garden. What they had in mind was something like a nineteen-thirties motor court.

Davis agreed. He instructed them to do something that would reflect regional traditions. But Taylor was not impressed by the local architecture of the Florida Panhandle. "It was very spare, a sort of impoverished version of classical architecture. I was worried that if we imitated it literally, the result would not be very appealing. We needed to dress it up." Taylor's family was originally from the Bahamas, and she used the islands' tropical architecture as a starting point. "I'm not a purist about style," she says. "I'm fundamentally a romantic. I imagined the houses as flamboyant cousins from the Caribbean who had come to visit their poor relations in Florida." The picturesque porches she designed were Bahamian; so were the pastel colors, although toned down for the harsher Florida light. Davis had a collection of old wooden brackets and pieces of decorative scrollwork that Taylor and Orr incorporated into the porches, which gave them a vaguely Victorian appearance. They added fanciful gates, gazebos, and benches in the garden. Davis named the little group Rosewalk, after his fiancé, Daryl Rose, to whom he gave one of the cottages as a wedding present.

Rosewalk was completed first, so it dominated the early media coverage of Seaside. It was what visitors remembered, and it was what the first builders copied. Although the code did not specify Victorian brackets and pastel colors, thanks to Rosewalk, these quickly became the hallmark of Seaside—and eventually of traditional neighborhood development itself.

What Orr & Taylor did at Rosewalk could be described as theming. Theming provides a coherent and instantly recognizable set of visual cues, to the home builders as the development is being created, and later to the people who live there. Theming, as David Brooks puts it, can "take something bland and give it personality and a sense of place."[4] Building according to a theme has a long architectural pedigree—just think of Marie Antoinette's hamlet at Versailles—but its modern roots are in the late nineteen twenties in Southern California. Knott's Berry Farm, outside Los Angeles, was an amusement park in the form of an Old West ghost town, with buildings that were moved from an actual abandoned town in Arizona.[5] In 1953, when Walt Disney was designing Disneyland in nearby Anaheim, he copied the idea but went further, building not only Frontierland but also Adventureland, Fantasyland, Tomorrowland, and Seaside's predecessor, a replica of an idealized nineteenth-century American small town, Main Street U.S.A.

Yet when Disney built a real town, it didn't use a theme. The planned community of Celebration was undertaken mainly as a way of developing a piece of real estate that adjoined its Walt Disney World theme park near Orlando, Florida.[6] Once the decision was taken to design a traditional-looking town, the planners, Robert A. M. Stern and Jaquelin T. Robertson, recognized that central Florida had an eclectic mix of architectural styles.[7] Instead of mandating one, they gave builders a choice: houses could be Classical, Colonial Revival, Coastal, Victorian, French, Mediterranean, or Craftsman. An elaborate pattern book spelled out the details of each style.[8]

While theming represents a desire to create a consistent and instantly recognizable sense of place, Celebration's eclecticism pragmatically accepted that architectural style is a matter of taste, and that urbanism has traditionally incorporated variety. I ask Jason Duckworth which approach he favors. He appreciates theming, he says, having learned about it from Davis, who doesn't like eclecticism. But since he's joined his father and become immersed in the business side of development, he has come to appreciate the advantage of offering buyers choices. He also recognizes that creating a themed community is not easy. "When the architecture calls attention to itself, it has to be done *very* well to be successful," he says. As a novice, Jason is uncomfortable with trying to do too much. "We're not trying to break new ground at New Daleville," he says, somewhat defensively.

I ask Mike DiGeronimo the same question. DiGeronimo, a slow-speaking, thoughtful man, is an architect and planner who works for Looney Ricks Kiss, known as LRK, a Memphis-based architectural firm that specializes in neotraditional communities and has offices in Nashville, Celebration, Rosemary Beach, Florida, and Princeton, New Jersey. Following its agreement with the township, Arcadia hired LRK to review the builders' house plans for New Daleville. DiGeronimo will be the "town architect." The function of a town architect, another Duany and Plater-Zyberk invention, is to help builders understand and implement design guidelines. DiGeronimo and I are sitting in a conference room in LRK's Princeton office, an old frame house just off Nassau Street. He comes down firmly on the side of eclecticism and agrees with Jason that a development with a theme is harder to build. It is also harder to market the houses, since home buyers like to have choices. To underline the importance of choice, he tells me about an LRK project in Michigan, where three builders offered houses in a variety of styles but one "preprogrammed" the lots, that is, he assigned specific styles to specific lots. He lost so many sales that after a few months he gave up and let people choose whatever they wanted.

DiGeronimo's approach to style is equally pragmatic. "We tend to discourage builders from trying to build specific historical styles," he says. "It usually doesn't work out since they don't know enough about the subject. We also think that often the money could be better spent elsewhere in the development, such as on landscaping or higher-quality materials." I ask him what are the most common stylistic mistakes builders make. "They try to put together different pieces that don't fit," he says. "They like this porch and that door and that bay window, but they don't know how to integrate them. They also often get the proportions and details wrong. So they end up spending money illogically, spreading the dollars around. We show them how to fix that."

So what will the New Daleville houses look like? According to DiGeronimo, whose firm is currently town architect for Celebration, Disney's eclectic approach won't work here. "We won't try to be stylistically accurate. Disney had the resources to develop a very detailed pattern book for builders to follow. High prices have attracted custom builders, who generally employ more skilled tradesmen. Also, Celebration is large enough to support a full-time town architect, while I'm only going to be at New Daleville half a day every few weeks." So what's the alternative, I ask him. "We'll focus on the general massing, the details, and the scale," he says. "You could call it a generic neotraditional approach."

I'm interested in his answer. This is neither theming nor eclecticism. At first glance, "generic neotraditional" sounds wishy-washy, but on reflection I realize that this is what American developers have always done. Stylistically consistent developments, such as Forest Hills Gardens or Seaside, are rare. So are developments where skillful architects have used specific styles. Even in Chestnut Hill, which has themed groups of houses, such as French Village, most of the houses that Woodward's architects built can't be categorized stylistically. They have stone walls and steep roofs, porches and dormers, but little, if any, historical detail. There may be a pediment over the front door, but that hardly makes them Classical. Most Americans would recognize them immediately—they are generic traditionals.

The Dream

One should not underestimate the importance of Levittown. It introduced the American public to modern production building and proved that standardization, mass production, and technical innovation could be successfully used to produce houses for a large market, not by manufacturing houses in factories but by intense standardization of the product itself.

Generic traditional houses are expressions of a common American taste. This preference is bound up in another national credo: the American Dream. The term acquired common currency in the nineteen thirties, thanks to *The Epic of America* by James Truslow Adams. "There is an 'American dream' of a better, richer and happier life for all citizens of every rank," the Pulitzer Prize–winning historian wrote.[1] However, for most people at that time, the dream of a better life did *not* include owning their own homes. Rapid urbanization had had a severe impact on the tradition of living in your own house. By 1930 fewer than half of American households were homeowners, and these lived chiefly in rural areas. City dwellers, except for a privileged few in garden suburbs, occupied rented tenements and flats.

During the Depression, the home ownership rate dropped to less than 44 percent. Then, thanks largely to postwar prosperity and the intervention of the federal government, it rebounded. Hoover's Federal Home Loan Bank Act of 1932 and Roosevelt's Federal Housing Administration stabilized the mortgage market and provided insurance for home mortgages as well as for housing construction loans. Since building had largely stopped during the Depression and the war, the demand for housing was huge. The question was how to meet it. Many people thought that they had the answer. Old New Dealers promoted government-built towns, but the bureaucracy was too slow in reacting to the accelerating demand. The followers of the European International Style proposed high-rise apartment towers. Buckminster Fuller unveiled the Dymaxion house, which was to be manufactured in an aircraft factory and resembled a flying saucer. The solution proved to be something entirely different and unplanned: mass-produced suburbs built by private developers or, as the housing historian Marc Weiss calls them, community builders.

"A community builder designs, engineers, finances, develops, and sells an urban environment using as the primary raw material rural, undeveloped land," writes Weiss.[2] The prototype for post-war community builders was Levitt & Sons. The firm built large planned communities—named Levittowns—in New York, Pennsylvania, and New Jersey. The first Levittown, begun in 1947 as two thousand rental houses on what had been potato fields in western Long Island, was part of Truman's Veterans' Emergency Housing Program. The popular demand for buying houses proved so strong, however, that the Levitts converted the rentals to ownership, acquired more land—ultimately 4,700 acres—and in only five years built 17,400 houses.

The selling price of a Levittown house was remarkably low: $7,500 ($48,000 in today's dollars). Returning GIs could become homeowners with nothing down and monthly payments of only $65.[3] The Levitts achieved dramatic economies—and a healthy profit of $1,000 per house—by reorganizing the construction process. The driving force behind this idea was William Levitt, the elder of the two sons. He had served in the Seabees, building barracks for enlisted men in Norfolk, Virginia, and he applied his wartime experience to the traditional world of wood-frame construction. Instead of building houses one at a time, he divided the construction process into twenty-six discrete steps, each performed by a separate team of workers, equipped with labor-saving devices such as power tools and paint sprayers. "One team would lay the slabs, another would do the framing, another the roofing and so on," he later recalled. "What it amounted to was a reversal of the Detroit assembly line. There, the car moved while the workers stayed at their stations. In the case of our houses, it was the workers who moved, doing the same jobs at different locations. To the best of my knowledge, no one had ever done that before."[4] To bypass unions, Levitt hired the workers as subcontractors. To speed up the work, he paid them not by the hour but according to the number of houses completed, and traded bad-weather days for Saturdays, Sundays, and holidays.[5] Thanks to such tactics, Levitt & Sons boasted that, at the height of construction, a house at Levittown was completed every eleven minutes.[6]

William's brother, Alfred, a self-taught architect, was responsible for design. In 1937, when he was twenty-five, he had taken a leave from the family business to follow daily the construction of a single-family house in Great Neck, Long Island. The architect was Frank Lloyd Wright, and the house was one of his so-called Usonians.[7] Wright had coined the term the year before to describe the prototype of the sort of house he visualized for Broadacre City. If he could not realize his urban vision, he could at least show people what a home in the City of Tomorrow would be like. Over the next two decades, he built more than a hundred such houses across the United States.

"The house of moderate cost is not only America's major architectural problem," Wright proclaimed, "but the problem most difficult for the major architects."[8] Of course, it wasn't too difficult for him. To reduce cost, he invented a highly simplified and modular method of wood construction. He eliminated the basement and the attic, and replaced the garage with a carport. He introduced a novel form of heating—under the floor. He made the kitchen a small work area and combined the living and dining rooms into a single space. He used polished concrete floors and exposed wood walls and ceilings—natural-looking as well as economical—and designed built-in furniture. Thanks to such innovations, he was able to build houses—beautiful houses—for as little as $5,500, at a time when his Fallingwater cost $166,000.[9]

The Great Neck house, one of the first Usonians, was larger than the houses that the Levitts would build; it cost $35,000 and took ten months to construct, but it had a powerful influence on young Alfred.[10] His first Levittown house was a 750-square-foot cottage, with two bedrooms, a living room, a bathroom, and a kitchen. The interior was small, but an unfinished attic had space for two extra bedrooms, and the sixty-by-one-hundred-foot lot left plenty of room for expansion. The model was called the Cape Cod, which conjures up a traditional image (as it was intended to), but the design incorporated several Usonian features, such as no basement and radiant heating in the floor. The Cape Cod was only the beginning. To attract buyers, the Levitts changed models every year, which allowed Alfred to introduce more innovations. These included open plans that combined kitchen, living, and dining spaces; central fireplaces; built-in televisions; and carports. Following Wright's example, Alfred planned the house on a two-foot grid that was laid out with string on the building site as a construction guide. He used modern materials, such as plywood instead of planks, and sheets of gypsum wallboard instead of hand-laid plaster. Thus the young developer became an unlikely conduit for disseminating Wright's ideas into the American mainstream.

Levittown was not just a large housing development, it was a planned community. "The veteran needed a roof over his head and instead of giving him just a roof we gave him certain amenities," recalled William Levitt in a 1983 interview. "We divided [the community] into sections and we put down schools, swimming pools, and a village green and necessity shopping centers, athletic fields, Little League diamonds. We wanted community living."[11] Levittown was advertised as a Garden Community, which was an explicit reference to the earlier garden suburbs.[12] As the city planner Alexander Garvin points out, it was a highly simplified version of Frederick Law Olmsted's Riverside, with curving streets, many trees (planted by the developer), and houses set well back from the sidewalk on large lots.[13] As in Riverside, front fences were prohibited, which left the lawns open, giving the impression of a continuous green landscape.

What set Levittown apart from previous residential developments was the number of houses, the speed with which they were built, and also their extreme architectural uniformity. Although buyers were offered relatively minor façade variations, as well as several colors, at any one time there was but a single basic house plan. This repetition reduced construction costs by enabling the work crews to repeat building operations efficiently and to use precut lumber and identical components.

Although low prices made the houses popular with buyers, Levittown had many critics. Modernist planners, influenced by European theories of urbanism, preached the virtues of multifamily housing. Architects, caught up in the International Style, were not impressed by Alfred's adaptation of Wright. Proponents of factory prefabrication and intricate building systems ignored William's organizational achievements; they couldn't believe that something as simple as Levittown could represent the future. The new mass-produced suburbs were the subject of several critical books. David Riesman's *The Lonely Crowd* and William H. Whyte's *The Organization Man*, which was based on his study of a suburb in Chicago, both maintained that a low-density suburban environment made people isolated and unhappy, as well as more conformist. This theme was echoed in fifties novels such as Sloan Wilson's best seller, *The Man in the Gray Flannel Suit,* and John Keats's satirical potboiler, *The Crack in the Picture Window.*

One sociologist questioned the popular wisdom. "I watched the growth of this mythology with misgivings," wrote Herbert J. Gans, who had worked under Riesman as a student and taught at the University of Pennsylvania, "for my observations in various new suburbs persuaded me neither that there was much change in people when they moved to the suburbs nor that the change which took place could be traced to the new environment."[14] In 1958 Gans began a firsthand study to determine exactly how people were influenced by moving from cities to suburbs. He and his wife bought a brand-new four-bedroom Cape Cod model in Levittown, New Jersey. After two years, and a further period of observation, he published his now-classic study, *The Levittowners.*

Gans pointedly described the difference in values between the critics of suburbs, who were largely intellectuals and city dwellers, and the people who lived in Levittown, who were generally blue-collar workers and low-level administrators. "[City planners] argue for suburban row house and apartment construction because they believe that the single-family house wastes land, encourages urban sprawl, increases commuting, makes for a physically monotonous community, and discourages an urbane way of living," he wrote. "Few actual or potential suburbanites share these attitudes, because they do not accept the business efficiency concept and the upper middle class, anti-suburban esthetic built into them."[15] Gans pointed out that, while the physical homogeneity of mass-produced housing came in for much criticism, upper-class nineteenth-century Georgian town houses (also built by developers) were likewise highly homogenous, and in any case, today nobody but the wealthy could afford custom-built houses.

Gans vigorously contested the view that the suburbs were imposed on unwilling consumers. "Levittown permits most of its residents to be what they want to be," he wrote, "to center their lives around the home and the family, to share leisure hours, and to participate in organizations that provide sociability and the opportunity to be of service to others."[16] He conceded that low-density suburbs were far from perfect, especially for teenagers and elderly people, who generally did not own cars. "It is easy to pro-pose community improvements and even utopian communities that will make Levittown seem obsolete," he wrote. "Yet I should like to emphasize once more that whatever its imperfections, Levittown is a good place to live."[17]

To see for myself, I visit a Levitt development in Bucks County, which is close to where I live. This was the second Levittown, started in 1951, just before Levittown, New Jersey. With more than seventeen thousand homes on six thousand acres, it was advertised as "The Best Planned Community in America." Alfred Levitt divided the site into what he called master blocks, about a mile square, bounded by parkways with limited access. Inside each block, he laid out three or four neighborhoods of four hundred houses, separated by local streets and landscape features. At the center of each block, no more than a five-minute walk from any house, was an area dedicated to community uses: parks, recreation centers, swimming pools, or schools. There were no commercial buildings. Instead of local village centers, as in the Long Island development, the Levitts built a mile-long commercial strip along the highway. "Thanks to the number of appliances in our house, the girls have three hours to kill every afternoon," said Alfred. "They want to find some excitement and they prefer to do even their grocery shopping in the main retail district."[18] Of course, the "girls" would drive to the supermarket; all the houses had two-car carports. The automobile-centered plan was Alfred Levitt's ver-sion of Broadacre City.

Driving along Levittown's winding streets today, it's hard to distinguish the original architectural features of the houses, since they are swathed in garages, annexes, and extensions that have been added during the last fifty years. The trees, planted by the

Levitts and now sixty feet tall, as well as the variety of individual gardens, likewise dispel any lingering impression of uniformity.

The Levitts were sensitive to criticisms of the standardized design of their houses, and for Levittown, Pennsylvania, Alfred created six different house models. The workhorse of the development was the Levittowner, which is his most Usonian design. It is what became popularly known as a ranch house. The Wrightian exterior, with its low-slung appearance, carport, high-silled bedroom windows, and large areas of glass, was distinctly untraditional. So was the wall material, eight-foot-tall striated sheets of asbestos cement, called Colorbestos and developed especially for the Levitts by the Johns Manville Corporation. Alfred designed an open plan with two bedrooms and a third so-called study-bedroom that was separated from the living space by a basswood screen that slid on a metal track. The three-way fireplace was visible from the kitchen, the dining area, and the living room. The layout was unusual, since the kitchen faced the street. This arrangement saved money on water and sewer connections, but it also oriented the living room to the garden, in true Usonian fashion. Each house included newfangled conveniences such as a kitchen exhaust fan, preassembled metal kitchen cabinets, recessed lighting, and built-in closet shelving. The kitchen came equipped with a GE refrigerator and stove, and a Bendix washing machine. It also contained the house's heating furnace, which was designed and built especially for the Levitts by York-Shipley, a large boiler manufacturer. The furnace was small enough to fit under the counter, its top doubling as a hot plate.

At $9,990 for one thousand square feet, the Levittowner was an even better deal than the original Cape Cod. The buyers were chiefly workers at U.S. Steel's mammoth plant, newly built nearby. When the model homes were unveiled, even though it was midwinter, fifty thousand visitors showed up the first week. Sales were so brisk that, by the time production started in the spring, houses were being built at the rate of 150 a week.[19] At the urging of local government officials, the Levitts offered a two-bedroom rental unit for sixty-five dollars a month. Since the monthly mortgage payment on a Levittowner was sixty dollars, there were few takers, and the so-called Budgeteer was soon discontinued.

Lewis Mumford contemptuously referred to the Levitts' developments as "instant slums," but there was nothing slumlike about Levittown, Pennsylvania, when it was built in 1951, nor is there today. The houses are well-maintained and the gardens carefully tended. According to the local real estate listings, an expanded Levittowner (the name is still used) sells for about $200,000. There are recreational vehicles parked in the driveways, a lively profusion of lawn ornaments, and many garden sheds and aboveground pools. This has remained a middle-class community, which would probably have pleased Alfred Levitt. "There's no point in trying to do something," he once said of his architectural work, "unless it can be handed out to the great masses of people as a cultural increase."[20]*

When Levittown, Pennsylvania, was being built, *House + Home* magazine praised the "progressive" designs, because "in the years ahead Levitt houses will be less out of date than old-fashioned houses built at the same time."[21] In fact, the houses have been greatly modified over the years. Asbestos has been covered with vinyl siding and brick, windows have sprouted shutters, large areas of Thermopane have been replaced by traditional bay windows, carports have been enclosed, and some of the ranch houses have grown second floors. The truth is that most of Alfred's Usonian features have fallen prey to changing fashions. Yet one should not underestimate the importance of Levittown. It introduced the American public to modern production building and proved that standardization, mass production, and technical innovation could

---

*After falling out with William in 1954, Alfred Levitt struck out on his own. He developed a thousand-unit apartment complex (today known as Le Havre) in Queens and a suburban subdivision in the Islip area, both innovative designs. He died in 1966, only fifty-four.

be successfully used to produce houses for a large market, not by manufacturing houses in factories but by intense standardization of the product itself, and rationalization of operations on the building site. It also showed that middle-class Americans were attracted to home ownership and suburban living no less than were their wealthier counterparts. Finally, it demonstrated how entrepreneurial efforts could create cheap, quick, lasting, and flexible housing.

Builders

A national builder sells the same four-bedroom house for $300,000 in upstate New York, $450,000 in Chester County, Pennsylvania, more than $750,000 in Prince Georges County, Maryland, and almost $1 million in Loudoun County, Virginia.

Vince Graham is one of the most successful neotraditional developers in the country. I met him about twelve years ago, in Beaufort, South Carolina. My wife and I had decided to go there for spring break, largely on the strength of having seen *The Big Chill*, which was filmed in the Lowcountry coastal town. Beaufort was settled in 1710 and long functioned as a port, but it rose to prominence as a summer retreat for antebellum planters. They built charming houses with raised verandas and deeply over-hanging roofs, set on meandering shaded streets that crisscrossed a small peninsula—Old Point—sticking out into the Beaufort River.

One afternoon, leafing through a local Chamber of Commerce brochure looking for a restaurant, I came upon a full-page adver-tisement for a real estate development. What caught my eye was the caption: THEY HAD IT RIGHT THE FIRST TIME. The development was called, rather cheekily I thought, Newpoint. I was intrigued.

According to the small map, it was on Lady's Island, just across the river.

Newpoint turned out to be a small neotraditional subdivision with small lots and back lanes. With its huge old trees and narrow streets, it really did resemble Old Point. At one end, near the water's edge, was a public park. The old-fashioned houses were an eclectic mixture of brick, wood siding, and stucco—an appealingly unpretentious neighborhood. After walking around, we stopped at the construction trailer that served as a sales office. That's where I met Graham, a soft-spoken native Atlantan, who looked much younger than his twenty-nine years. Newpoint was his first real estate venture, he told me. It was a bootstrap operation. He got preliminary approval to build, presold a few lots to friends and relatives, and then used the money as down payment on the land. "At one point I had to add a check written on my credit card," he told me. Yet after two years, only 30 of the 126 lots remained unsold, and he and his partner had almost finished paying off the land.

At the time of my visit to Beaufort, neotraditional development was in its infancy. As far as I knew, Newpoint was the first year-round neotraditional development to be a resounding success, although Graham seemed unaware of this distinction. I asked him how he had arrived at his concept. He told me that, as a freshly minted University of Virginia graduate in economics, he had come to the Lowcountry five years earlier to work as a project manager on a three-thousand-acre golf-course community on nearby Spring Island. He lived in Beaufort. Spring Island was to consist of single-family houses on extremely large plots, but Graham thought that a village type of development modeled on Old Point would make a good addition. His employers were not convinced. "That's when I showed my proposal to Charles Fraser," Graham told me.

Charles E. Fraser (who died in 2002) is a real estate legend. He pioneered the concept of the resort community at Sea Pines Plantation on Hilton Head Island, South Carolina, where between 1956 and 1975 he built more than three thousand homes. He also exerted a significant influence on the real estate industry, since he not only popularized concepts such as modern deed covenants, architectural review boards, and environmental planning but also trained a generation of outstanding residential developers. When Graham knew him, Fraser was retired though still working as a consultant, and he took the younger man under his wing. "You need to check out a place called Seaside," he told him.

Graham didn't actually go to Seaside, but he did read about it. His approach at Newpoint was simplicity itself: he found a street in Old Point he liked, measured it, and had his planner duplicate the dimensions. His other models were Savannah and Charleston, both within an hour's drive of Beaufort. He could not have chosen better, for these two cities are the Mecca and Medina of American Colonial urbanism. Savannah was laid out in 1733 by James Oglethorpe, as a repetitive system of regular squares and tree-lined avenues that produced the "supreme achievement of American gridiron planning," according to Jaquelin Robertson.[1] Charleston, older and with a more conventional plan, is characterized by exceptional architecture, both civic and domestic.

Graham first realized that he had done something special at Newpoint when Fraser showed up with a group of Walt Disney executives, including Peter Rummell, the head of Disney's real estate development division. Rummell, also a Fraser protégé, was planning what would become the new town of Celebration. A year later, two busloads of Celebration builders, managers, and architects (including Robert A. M. Stern and Jaquelin Robertson) descended on Newpoint. "These people had come all that way to see a little project done by us country boys," said Graham.

After completing Newpoint, Graham was ready to take on something bigger. He found a large parcel of land in Mount Pleasant, across the Cooper River from Charleston. After a contentious two-year struggle—not long by Pennsylvania standards but unusually protracted for South Carolina—to have the site rezoned for smaller lots and higher density, he finally got planning approval in 1997. I'On, named after the eighteenth-century owner of the plantation that originally occupied the site, is almost 250 acres divided into 750 home lots and includes two lakes, a small village center, a clubhouse, and shoreline walking trails. The development has been a great success. In 2000, *Professional Builder* magazine named it a Best Community; the following year the National Association of Home Builders gave I'On its Platinum Award for the best smart growth community in the nation, and the Congress for the New Urbanism awarded the development a Charter Award for outstanding urban design.

A large part of I'On's success is attributable to the high quality of its architecture. Graham doesn't believe in complicated or restrictive design guidelines. "My emphasis is on enabling good design, rather than preventing bad design," he says. He hired a local preservationist who was knowledgeable about the traditional architecture of Charleston. "I call him a design coordinator. His job is to work with homeowners and builders to ensure that the design of the houses reflects Lowcountry traditions," he explains. Graham's approach to builders is also unusual. The time-consuming experience of explaining his ideas to different builders at Newpoint had convinced him to try a different method. Before starting I'On, he surveyed several hundred small local builders and invited ten to participate in what he called the I'On Guild. Only guild members—there are now eighteen—can build at I'On. There is a formal investiture ceremony for new members, and Graham hosts workshops, dinners, and lectures—I once gave a talk to the guild—all with the goal of developing what he calls "a culture of the right way to do things." "We build the public realm and the builders build the private realm," Graham says. "We treat them as partners, and we make sure that when we make money, they make money."

The builders at I'On are local custom builders, that is, individ-

LAST HARVEST

ual contractors who build as few as one or two houses a year—or
as many as thirty. Custom builders typically build either from
architects' designs or from their own plans, which they modify to
suit a buyer's requirements. Such modifications drive up the
price, so custom builders produce only a tiny fraction of all
houses built annually in the United States. The overwhelming
majority are built by so-called production builders, who build several hundred houses a year and achieve economies of scale that
enable them to sell at lower prices. They use stock plans and
offer buyers a limited set of options, such as alternative façade
treatments or a choice of interior finishes. Large production
builders sometimes acquire raw land, but they also buy improved
lots from developers. After I'On's success, several production
builders approached Graham, but he didn't invite them to join the
guild. "I find that custom builders are faster on their feet," he says.
"They can learn new ways of doing things. The production
builders have a different mind-set. They tend to be more rigid, and
while they offer a cheaper house, they can't achieve the quality I'm
looking for. And if I allowed production builders in, they would
starve out the custom builders."

Graham's attitude toward production builders is consistent
with his conservative approach to development. A conventional
development scenario entails buying the entire site, building the
infrastructure, then selling the lots as fast as possible to repay the
bank loan. Production builders are an essential ingredient in this
process, since they alone have the resources to build—and market—large numbers of houses. Like Robert Davis at Seaside,
Graham has opted to go slow. This is partly because, like Davis, he
is highly risk-averse. "We're a small family company—my father
and brother are investors," he tells me. "We simply couldn't
afford to buy the entire site and put in all the infrastructure at
once. It would also leave us too exposed in case of an economic
downturn." Graham has an agreement with the landowner that
allows him to buy the land parcel by parcel, spread out over a ten-

year period. He sells only seventy lots a year. As one parcel sells out, the cash flow pays for purchasing the next, and so on. "This also allows a slower rate of building, which increases the quality," he says.

I ask Graham how going slow compares in profitability with a more conventional approach. "Initially it's probably more expensive. However, since we're selling the lots slowly, we're able to benefit directly from increases in land prices," he explains. "So I think that in the end, we're ahead." At I'On, the increases have been significant. The first lot sold in 1998 for about $45,000. Six years later, with the project still not finished, the cheapest lots are selling for $130,000, and choice lots are several times that price, although Graham is still paying the same buying price. "This year we met our sales quota by June," he says. "We raised the prices of the last few lots, and they still sold. Can you believe that our current profit margin on a lot is sixty-four percent?" In terms of real estate development, this country boy has hit the jackpot.

When I tell Joe Duckworth about my visit to I'On, praising the quality of the architecture, he becomes a bit defensive. "I'On is a handcrafted, high-end proposition," he says. "The reason that Graham is able to do what he does is that he is working with custom builders. In our area, custom builders only build houses over one million dollars. We will definitely be working with production builders, and probably with national builders."

Home building is a $50-billion-a-year industry, but despite the early success of the Levitts, it has been slow to consolidate and is still made up of numerous small firms; according to the latest figures, 94 percent of the 100,000 U.S. builders have fewer than ten employees.[2] At the other end of the spectrum, the 200 largest production builders in the United States accounted for 40 percent of all new houses in 2005.[3] Fully half of this production was dominated by only ten large, publicly owned corporations, the so-called nationals.[4] National builders are very large indeed; the largest, Texas-based D. R. Horton, builds more than fifty thousand homes a year. The nationals' chief competitive advantage is financial: as publicly owned companies, they have access to cheaper capital. In addition, they build in multiple markets, which makes them less vulnerable to local slowdowns. National builders buy materials and components in bulk, operate their own manufacturing plants, and sometimes own their own building supply and mortgage companies.

Unlike the manufacturers of refrigerators or automobiles, national home builders do not have standard prices. A national builder sells the same four-bedroom house for $300,000 in upstate New York, $450,000 in Chester County, Pennsylvania, more than $750,000 in Prince Georges County, Maryland, and almost $1 million in Loudoun County, Virginia.[5] The difference is not the construction cost (which in all cases is in the neighborhood of $90,000), but the cost—that is, the availability—of permitted land and the strength of demand. The profit margins vary, too, hot markets subsidizing lackluster ones.

The houses produced by national builders are not significantly different from those of their smaller competitors, except in one important aspect: consumer satisfaction. In 2003 the J. D. Power and Associates annual national survey ranked national builders first for customer satisfaction in twenty out of the twenty-five largest home-building markets.[6] The most satisfied customers were in Las Vegas, Austin, and Phoenix, the least satisfied, scoring well below the national average, were in Seattle, Washington, D.C., and Philadelphia, where the score had actually fallen since the previous survey. "Philadelphia builders are not as quality-conscious as the builders in more competitive markets," one builder was quoted as saying.[7] Jason Duckworth agrees. "Supply constraints on land have dampened competition in our region," he says. "While I think that we're going to do a nice job at New Daleville, it's really been self-motivated. The market has had little to do with it. The builders are so desperate for lots that they'd take most anything. In contrast, I think that builders in Las Vegas are very aware of competitive differentiation and value design more."

Although some national builders, such as Toll Brothers, acquire raw land and function like traditional merchant builders, most buy permitted lots from developers. In regions where the supply of permitted land is limited, it's not unusual to find them taking on as few as twenty or thirty lots at a time. Joe Duckworth hopes that he can interest a national builder in New Daleville. He likes the nationals, he says, because they have the financial resources to do deals and to carry them through, which is not something smaller production builders can do. He recognizes that it will not be easy to achieve architectural quality with a national builder. All production builders, largely in response to market demand, concentrate on the interiors rather than on the exteriors. At New Daleville, they will be asked to pay more attention to exterior details and materials. "To do a successful project, we've got to get the builders to do the best job they've ever done, and then some," he says. "That's why we hired LRK. They've developed a good method for communicating with builders without having them go batty on us."

19

A Compromise

Why it takes two years to get approval for a development project.

Jason Duckworth asks Mike DiGeronimo, as town architect of New Daleville, to lead a design workshop for a group of builders who have shown interest in the development. During the workshop, DiGeronimo explains the design review process, which will involve several meetings to go over sketches, schematic designs, and construction documents. He shows the builders how a house should respond to particular site conditions, such as corner lots. He uses photographs of do's and don'ts: shutters that are not properly sized for the window, for example, or poorly designed entrances. He explains the right way to detail a column, and to proportion a dormer. He demonstrates how a complicated façade, with too many architectural features, can be simplified and improved. It's all rather straightforward, remedial teaching, really.

Once the final plan for New Daleville is approved by the township, Arcadia will invite the builders to submit bids for the lots, but for now, Jason just wants to create interest in the project. New Daleville will not have a guild, but he wants to establish a relationship with the builders. "It's worked out well so far," he says. "Not only has the workshop attracted half a dozen good builders, including three nationals, but Mike's presentation has set high expectations for quality. Thanks to the reputation of LRK, and the design guidelines, the builders now see that New Daleville could be higher-end than they first imagined."

During the workshop, several of the builders ask about the design guidelines. Who will be in charge of reviewing their plans, they want to know, the developer or the township? In other words, how bureaucratic is this process going to be? Jason assures them that design questions will be handled in a timely manner, but the truth is that the debate about the guidelines is far from settled. "One of the reasons that we brought in LRK was to try and convince the township, especially Tim Cassidy, to back off on their demands," he says. "Unfortunately, it hasn't worked. There are still about a dozen topics that Tim wants to see covered in the guidelines that I consider a matter of subjective taste. Things like proportions and window trim. He also wants the township to have an active role in reviewing house designs before they are built, which is something I'm totally against."

Jason is cautious about telling the builders what to do. "The trick is to identify those guidelines that the builders see as adding value to their houses," his father instructed him. "Anything that we ask them to do that the market doesn't value means that they will just decrease the amount that they are prepared to pay us." Since Jason has been unable to decide exactly what will add value to the house, the guidelines he has written leave as much as possible to the discretion of LRK's review process. The neotraditional ordinance requires that a quarter of the houses at New Daleville have porches, but their location is not specified. Builders are not restricted as to exterior materials; they can use brick, stone, synthetic stone, stucco, wood siding, fiber concrete siding, or vinyl siding—almost anything. The guidelines forbid glass block, jumbo-size brick, and double-height glazed entrances (those monumental, out-of-scale front doors so beloved by builders), but those prohibitions will hardly ensure that houses will be well-designed. Tom Comitta, who is reviewing the guidelines on behalf of the township, has suggested prohibiting fake window shutters, faux architectural elements, and plastic porch columns and railings. "We agree with your intent," Jason writes in response to his suggestion, "however, we believe that this is a complex and subjective matter better enforced by LRK in the design review process." In other words, trust us.

It's instructive to compare the New Daleville guidelines with Vince Graham's Newpoint code, which he wrote himself. He got a copy of the Seaside code and followed it closely. Since Newpoint was not a luxury project—the original lots sold for only $20,000—and the market for such a development was untested in South Carolina, he simplified the rules. He did not require picket fences, for example, and he gave builders a choice of several materials for roofing, including conventional asphalt shingles. The Newpoint code was succinct—only two pages—and made clear and direct demands on the builders. One rule required that all houses have front porches at least half the width of the façade. At Newpoint, some of the builders followed the minimum requirement, others extended porches across whole façades, and a few even built two-level verandas.

Local regulations did not require the Newpoint architectural guidelines to be part of a municipal ordinance—they were attached to the homeowner agreement—so Graham was free to make changes to the code and grant waivers in exceptional cases. In contrast, at New Daleville, the guidelines will be part of the township ordinance, so future changes or exceptions will have to be formally approved by the township. Paradoxically, to ensure that he has discretionary control over design, Jason has had to keep design out of the guidelines as much as possible.

Until the New Daleville architectural guidelines are formally ratified, the township will not give its final approval to the project. Under normal circumstances, rural township officials don't involve themselves in architectural questions, but Tim Cassidy is hardly a typical official. Not only is he an architect but he's opinionated and occasionally even obstreperous. He has also spent a lot of time thinking about the architecture of Chester County. He says that if I want to understand his ideas, I should read his doctoral dissertation, although he warns me that it contains a lot of "architectural mumbo jumbo." It turns out to be a study of the building traditions of the Brandywine Valley. Cassidy concludes that the area does not have a distinctive regional style; instead, its architecture is "regional" because of its "engagement with the region, not as a result of its formal characteristics."[1] This subtle but unsentimental view explains why he accepts a generic approach to domestic design—his own house could be described as generic traditional. Although he is no purist, Cassidy takes issue with the way contemporary builders self-consciously dress up their houses to make them look old-fashioned. "I object to the use of applied plastic moldings, which are a weak attempt to mimic traditional forms of ornamentation," he says. "They are usually out of scale and are not integrated into the overall material assembly, simply screwed or glued to the surface after the fact. I think that the key to success with projects that have tight construction budgets and aggressive building programs is to keep the details as simple as possible."

Cassidy believes that the New Daleville guidelines in their current form will not prohibit the kind of banal suburban houses that have been built in his backyard. "The onus is on the developer, that is, Jason and company, to come up with a vision for what they want New Daleville to be," he says. "I've tried not to influence them stylistically, but they have to set some rules for themselves. When we went on that tour of developments, I would point out something in one project, and he would say, 'Great, we could do that.' Then we would see something different in another development, and he liked that, too. I think that Jason doesn't really know what he wants. He wants the builders to tell him what *they* want."

Cassidy has tried to convince Jason to make the guidelines more rigorous. He finally has come around to the realization that this is not going to happen, and that the quality of the project will depend on the review process. That's all right with him. "I've talked with Mike DiGeronimo, and we agree on most things. I have confidence in his judgment," Cassidy says. "But I'm concerned that if he makes recommendations that cost too much, or slow things down, Jason and Dave will just ignore him." As an additional guarantee, Cassidy has suggested that the builders' plans should get a final "cursory review" by a third party, an architect working on behalf of the township.

Jason doesn't like this idea at all. "We have a big problem with the township reviewing the designs," he tells me. "A lot of municipal consultants see their job as dragging out the process and extorting money from the developer." He quickly adds that, on the whole, Londonderry has been very cooperative. "We have good communications with the township staff, and it's definitely the best township to work with of all our projects. But what if the political winds were to change, and an anti-growth group of supervisors came into power in a year or two, when we were halfway through the development? They could delay approval of a design modification to slow up or even stop the project. I'm not being paranoid. There's a subdivision in Chester County where builders are refusing to buy lots because it takes three to six months to get a building permit. I've been told that the township is doing this on purpose, to discourage development."

It is now more than eight months since the developers brought the architectural guidelines before the planning commission. The two sides have been going round and round in circles, their positions unchanged: Arcadia still wants the freedom to do anything it wants, and Cassidy still doesn't entirely trust the developers to do the right thing. It's a standoff.

A month later, I ask Cassidy if there has been any progress on the guidelines. "It's not settled yet," he says. "Everything's still up in the air." Well, not quite. The previous week, while Cassidy and his family were on vacation in Maine, the question of the design guidelines was more or less resolved. Bob Harsch, the township engineer, invited Dave Della Porta to discuss the outstanding issues pending final approval of the plans. They breezed through all the issues, which included sewage and water supply, traffic, the homeowner agreement, the transfer of the park to the township, and the financial assurances that the developers would have to provide in the form of an escrow account. There was only one sticking point, the design guidelines. Cassidy's suggestion of having a township consultant look at the builders' plans after LRK's review was brought up. Della Porta asked what would happen if there was a disagreement. Who would be the arbiter? Would they bring in yet another consultant? "The township understood that this process didn't make any sense," he told me later. "The clincher came when I told them that since we had already hired LRK to work with the builders, we didn't see why we should have to pay for a second review." His message was clear. Consultants are generally paid out of the developers' application fees, but if the township wanted architectural advice, *it* would have to foot the bill.

Della Porta could see that, while the township didn't want to be involved in reviewing house design, neither did it want to ignore Cassidy. So he offered to incorporate the latter's concerns in a handbook that would be part of the homeowner agreement. This would make them binding on the builders, although not subject to the township's review. He also suggested that the township could look at a draft of the handbook, although he was careful to specify that this review did not constitute approval. The design guidelines would remain as they were, and there would be no external architectural reviewer. It was agreed that this arrangement would be presented to the planning commission at its next meeting.

"I think Arcadia may be exploiting the summer apathy of the township leadership," says Cassidy of the tentative agreement. "They don't seem to want to trouble themselves with details." The truth is that nobody at the township, other than Cassidy, has strong feelings about architectural design. Unless he volunteers to review the builders' plans himself, which he has already said that he is unwilling to do, it looks as if he will have to accept the compromise. In fact, Cassidy says that he has no objection to leaving the guidelines vague and putting the details in a handbook. His problem is that, as far as he knows, such a handbook doesn't yet exist. "New Daleville seems to be Arcadia's first project, so they can't even show us a comparable document. But I want to see *something*, before I approve this," he says.

When the planning commission meets to grant final approval to New Daleville, Tom Comitta describes the new agreement and says that the developers have agreed to address Cassidy's concerns in a handbook. Della Porta, who is representing the developers, stresses that the presence of LRK, a nationally recognized firm, will ensure high quality in the development. Cassidy again argues for an outside consultant to review the house designs on behalf of the township. He maintains that this does not have to be a lengthy procedure. "All you need to do is to specify a range of details," he says. "After all, there aren't going to be a hundred different houses to look at, only a few models."

The other members of the planning commission have been silent up to now. One says that when he had some work done on his house, he found a carpenter whose work he liked, gave him a general idea of what he wanted, and then left him to it. Has the world become so vicious that we can't trust anybody? A man at the end of the table adds that he has confidence in LRK's ability to ensure that New Daleville will turn out well. The planning commission members don't seem to share Cassidy's skepticism. They've been listening to New Daleville presentations for more than two years, and if the developers are not exactly neighbors, neither are they complete strangers. They've played by the rules, and they deserve some consideration.

Della Porta, who senses that he has the support of most of the commissioners, now makes a suggestion. Arcadia will hire LRK to write the handbook, and what's more, he'll show the handbook to the supervisors before they meet to vote on final approval next month. Martha Detering, the new township supervisor, is in the audience and says that having the supervisors review the handbook is a good idea and should be written into the conditions. Arcadia will have to spend some money to hire LRK—Jason later estimates that it may cost as much as $20,000—but in return the township will not interfere in the design of the houses. The motion is put, and the members raise their hands in favor. "You voting for this?" Richard Henryson, the chair, asks Cassidy. He nods in agreement.

The final step in the permitting process is for the supervisors to give their approval. It's been seven months since they gave preliminary approval and a full two years since they passed the traditional neighborhood development ordinance, so this vote will be a milestone. I show up on time for the supervisors' meeting, but no one from Arcadia is there. For a moment I think I've mixed up the date, but at the last minute, Christy Flynn arrives; it turns out that Jason and Della Porta are tied up with other projects. The meeting starts at seven-thirty, as usual, but it's a long agenda, and New Daleville doesn't come up until nine o'clock. LRK has delivered copies of the new handbook—which seems to me suspiciously similar to the document Jason prepared months ago—but the supervisors appear satisfied. Nevertheless, there is a forty-five-minute discussion of the approval conditions, one by one. The details are picayune: the township wants to be sure that the homeowner association will be responsible for maintaining and replacing street trees; one of the supervisors thinks that plastic play sets shouldn't be permitted; how, exactly, will parking be regulated on the streets, asks another. Flynn handles the answers adroitly.

The unanimous vote of approval, when it comes, is an anticlimax. Just before the vote, Martha Detering makes the unexpected suggestion that the developers consider a different name. According to her, many local people don't like the name New Daleville. It's unclear exactly why. Couldn't it be called simply Daleville, she asks. This is the exact opposite of what the township wanted a year ago. Flynn quickly answers that Arcadia has no objection to changing the name, as long as it is done before the marketing starts. As has happened so often during these public meetings, the issue is left unresolved. (In fact, the name will stay the same.) But it is a sign of the township's growing identification with the project that the question of naming should even arise. That's very different from hiding the development behind berms.

PART THREE

New Daleville, June 2006

20

Trade-offs

How Wall Street causes national builders to pay more for building lots.

J oe Duckworth had described the development business as consisting of two distinct steps: spending money buying and subdividing land, and making money selling the lots. With the final plan approval in hand, he now has a "permitted" project, and it's time to move to step two: auctioning the lots to builders. Dave Della Porta is in charge of the bidding process. "Several months ago we contacted about twenty builders we thought might be interested in New Daleville," he says. "In addition to regional and national production builders, we included several local custom builders, to raise the quality of the project. Unfortunately, though we had seven or eight serious expressions of interest, they were only from the larger builders. There have been so few neotraditional developments in this area that the small builders are cautious and not willing to take the plunge." After the builders' workshop, the list is winnowed down to six—three regional builders and three nationals. They are invited to submit bids for the lots.

According to Della Porta's latest calculations, site improvements, consultants' fees, extra payments to the township, interest on the bank loan, as well as the price of the land, have pushed the cost up to $96,000 per lot. The bids of the three national builders exceed $110,000 per lot. This is considerably more than the bids of two of the regional builders. The third regional builder has bid slightly higher, but Della Porta is not convinced that he can deliver the quality required and eliminates him, too. Della Porta was hoping to have at least one regional builder for the sake of variety. "It's disappointing," he says. "The low bids reflect a lack of experience with the neotraditional concept, as well as concern with the location. New Daleville is at the edge of where development is currently taking place in southern Chester County. It's not close to employment centers, and there's still plenty of open land available in the area. All things considered, I would call it a C-plus or a B location."

Della Porta gives the three nationals a deadline to submit their "best final offer." The first reply comes from K. Hovnanian Homes. Hovnanian has built a neotraditional development in New Jersey but has not done any projects in Chester County and is keen to gain a foothold. The bid is $125,000 per lot. NV Homes and Ryan Homes are represented by their parent company, NVR, whom Della Porta informs of the new offer. "This is where the market is," he tells NVR. Reluctantly, the company agrees to match Hovnanian's price.

"NVR and Hovnanian have each offered to take all the lots, but we definitely want more than one builder on the project," says Della Porta. "The developers of the successful TNDs we've visited have all stressed the importance of variety." Della Porta, who's a builder himself, is worried about the design of the Hovnanian houses that he's seen, but the company has agreed to work closely with LRK. He divides the lots, forty-five to Hovnanian and forty each to NV and Ryan.

Della Porta explains why the nationals are able to pay more for the lots than the smaller, regional builders. "In a deal like this, a small production builder typically buys the lots all at once. This doesn't appeal to national builders, since Wall Street doesn't like them to carry unused land on their balance sheets. So we've offered them what is called a rolling option. They each make a half-million-dollar deposit, which is credited to lot sales on a pro rata basis. They are obligated to buy at least four lots per quarter, although there is an annual five percent escalation clause, which encourages them to take more. They have the option of pulling out of the project at any time. In return, they are able to pay a premium price."

"I've updated the project pro forma based on the latest costs and the new lot sale price," Della Porta e-mails Duckworth, "but I'm not going to show it to you because it looks too good." New Daleville now has a very healthy rate of return of 25 percent. "The hundred and twenty-five-thousand-dollar price is an astonishing number," says Jason, who is delighted with the outcome. "We can't really take credit for it. It's not the quality of our project, it's the layers of regulations that have shrunk supply and turned permitted land into a scarce commodity. But I tend to be the skeptic around here, so I won't really believe this price until we have the contracts signed." The bids are firm, but the contract will be signed only after a due diligence period, which gives the builders ninety days to verify the state of the various permits, study site issues such as grading, determine which of their models will fit on the specific lots they are buying, and have further discussions with LRK to pin down the exact cost of adhering to the design guidelines.

Everything is falling into place. The supervisors have voted to make the New Daleville guidelines part of the TND ordinance, and following several weeks of negotiation, the Department of Environmental Protection has given the nod to the sewage plan. There are only a couple of permits outstanding. "We thought we had an agreement with the township concerning off-site road improvements," says Duckworth, "but the state Department of Transportation has been pressing them to ask us for more money."

Until the agreement with Londonderry is signed, the state will not issue the so-called highway occupancy permit, which is required whenever a new access road is connected to a highway (New Daleville has three such connections). The delay means that Arcadia will have to get a temporary permit in order to start construction. This sort of thing is done all the time, Duckworth assures me.

A more important issue concerns water supply. Because there is no municipal water system in Londonderry, residential subdivisions are supplied by individual wells. However, since a recent drought, when many local wells ran dry, residents have become resistant to anything that threatens to lower the water table. For this reason, the developer of Honeycroft Village, which will have four hundred houses, has proposed to bring piped water into the township. The new three-mile, ten-inch-diameter pipeline will be built by the Chester Water Authority, whose source of supply is the Susquehanna River. Since the pipeline passes in front of New Daleville, it can service that project, too. Duckworth knew about the possibility of municipal water from the beginning, and it figured strongly in his calculations. The pipeline will supply another subdivision in the area on the way to Honeycroft, so his share will be about a quarter of the $2.4 million construction cost.

Chester Water Authority is ready to start laying pipe, but a snag develops. To reach Londonderry, the pipeline must cross a township immediately to the south, and a newly elected supervisor there has objected to the construction, on the grounds that the availability of piped water will encourage local development. According to the lawyers for Chester Water and Honeycroft, as well as Arcadia's lawyer, Marc Kaplin, the township can't block construction of the pipeline. "The pipeline is actually on the Pennsylvania Department of Transportation right-of-way," says Jason, "so we don't think that they have any legal grounds for objecting. There won't even be any outlets in the township. But Chester Water is getting spooked, and the project is being delayed."

Such last-minute delays are not unusual, and this one would be of little account except that something else has come up. Arcadia's option on the site expires in two months, and Dr. Wrigley's attorney informs the company that his client won't grant any more extensions. "He's figured out that land values have gone up, and he could get a lot more for his property than our original offer," says Duckworth. "In an ideal world we would ask for another extension, since we're not quite ready to go ahead, but as it stands we'll have to buy the land now." Acquiring the land before all the permits are in sounds risky to me, but Duckworth is sanguine. "This project has cost me about eight or nine hundred thousand so far. We don't quite have all the permits, but with the strong builders' bids, and with my track record, our bank is willing to advance us one and a half million to buy the land. It's a term loan for three years, with a one-year extension. If things go bad, we can still sell the land and pay back the bank. Not ideal, but not the end of the world."

I talk to Duckworth again the day after the closing. According to him, a sale like this has two aspects. "The business transaction is pretty straightforward. The second part involves selling a family legacy, which is not necessarily rational and can be an emotional experience. We had some preclosing jitters yesterday."

The agreement between Arcadia and the seller is now more than two and a half years old, and Duckworth estimates that in the interim the land has doubled in value. "Naturally, the landowner feels that he should get more, and naturally we feel that a deal is a deal. Usually we would prevail, except that there was a slipup. When our engineers laid out the project, they created a drainage easement that encroaches on Wrigley's five acres. Normally they would never have done this, but they probably thought it was all part of the same project. We weren't paying attention, and we didn't catch it. Since the seller approved all the drawings, we feel we're in a strong position, but technically it's a violation. To avoid litigation and keep things moving, we've agreed to pay an extra hundred thousand dollars for the easement. I try to think of it as eight hundred dollars off the price of each lot."

In January 2005 Arcadia organizes a day trip for the three builders to Florida, to visit Celebration. Disney's planned community is probably the best executed of the neotraditional projects, and Jason wants the builders to understand that they should be aiming high. Since LRK is also providing the town architect for Celebration, Mike DiGeronimo will head the tour.

Celebration, begun ten years ago, has grown to more than ten thousand residents. Although it's an overcast day, the place looks attractive. The recently completed portions have the slightly antiseptic appearance of all new master-planned communities, but the older neighborhoods have aged nicely, with mature trees and established landscaping. The landscape features are important, since Robert A. M. Stern and Jaquelin Robertson, the planners of Celebration, modeled the community on a classic garden suburb, with many small greens and parks.[1] The houses are in a variety of styles. The Spanish Colonial villas look particularly at home, as do the Lowcountry bungalows, with their deep verandas. By contrast, the Georgian town houses, which would be suitable for a London square, look odd shrouded in palmettos.

The tour focuses on a neighborhood that is being developed by a large production builder. DiGeronimo points out the simplicity of the architecture. He talks about one-sided houses and how corner lots need to address both streets. What makes a strong impression on the visiting builders is that these rather simple-looking houses on small lots are selling briskly. The group tours a model home, a four-bedroom Craftsman-style bungalow with a tiny back porch, that goes for half a million dollars. The Orlando real estate market is booming, and house prices have doubled in the last five years.

After lunch at the Celebration clubhouse, the builders discuss how they will differentiate their products—to home builders, houses are always "products." Ryan will sell the least expensive houses, NV the most expensive, and Hovnanian will be in between. They discuss logistics. Having three different contractors on the site at the same time will require close coordination of the various trades. The builders agree that it's important to have their lots in separate areas, to minimize friction between the work crews. "It's easy for us to agree around this table," says Brad Haber of Hovnanian, "but our guys in the field are pretty rough, and these lots are small. There's not much room for getting materials and equipment in and out."

I talk to Haber on the plane. He's been with Hovnanian for two years, in charge of land acquisition for the Delaware Valley division. I ask him about New Daleville. "It's only forty-five houses, but it's a chance for us to learn about neotraditional development," he says. "And it's such a small project that if it doesn't succeed it won't be a disaster." He would have preferred to take all the lots, rather than having three builders on the site, since the price range will be small and the competition will be fierce. He's also conscious of being squeezed between NV and Ryan, who are basically part of the same company. Haber describes the different business strategies of NVR and Hovnanian. "NVR are the top performer of all the publicly held national builders, so they have to keep delivering steady growth. They tend to buy land anywhere and then figure out what to build on it. Our strategy is to produce a variety of products across the whole market. In metro Philadelphia, for example, we've got single-family homes ranging from the mid–two hundreds all the way up to the mid–seven hundreds. With such variety, we tend to be very analytical."

A month later, I ask Jason about the final negotiations with the builders. He tells me that NV and Ryan have signed contracts and are working with DiGeronimo to finalize their house plans but that Haber has asked for an extension to the due diligence period. "I'm surprised, because K. Hov. came in with the strongest bid. I hope they're just being careful," he says. A week later Hovnanian pulls out of the project.

"The decision was not made lightly," says Haber when I ask him about the change of heart. "We were comfortable with our original house price estimate of three hundred and twenty thousand dollars, but when we looked more closely at the project, we realized that it needed to be closer to three hundred and eighty thousand. In this area, that's sixty thousand dollars above the market price for a similar house on a one-acre lot. Is this premium too big? We just don't know. Hovnanian has done neotraditional communities in other parts of the country, but not in Chester County. So we're not sure how strong the demand for this kind of house on a small lot will be."

According to Haber, the decision to pull out was not chiefly economic. "We strategized the following scenario. Let's say that we build two model homes and open up for sale. After an initial spurt of interest, sales slow down considerably; it turns out to be hard to sell houses on tiny lots in the middle of wide-open countryside. After a year and sixteen very slow sales, we decide to walk away, which our contract allows us to do. The economic cost is not large. We spend about a billion dollars a year on land acquisition in the Northeast, so in that context the loss is actually insignificant. But walking away at that late stage will seriously damage the project and hence our relationship with Arcadia, which is something we don't want to do. I've known Joe for twenty years. He's a visionary, plus he's very entrepreneurial, so he's someone I want to do business with in the future. Jeopardizing that relationship isn't worth the risk, so we're bowing out now. I'd like to say gracefully, but it's never graceful."

Joe Duckworth is unruffled. "I think Brad just couldn't convince his bosses that this was a worthwhile project. These things happen. We're not worried about the extra lots. We'll keep five ourselves, which we'll sell later, and divide the rest between NV and Ryan. We are slightly concerned about having sufficient variety with only two builders, but LRK should be able to take care of that."

But perhaps he should be worried. The departure of Hovnanian has radically altered the dynamics of New Daleville. Although NV Homes and Ryan market houses under independent brands, they are divisions of the same company. For all practical purposes, the development now has a single builder, a single very *large* builder. In 2004 NVR, with 4,500 employees, had combined revenues of $4.3 billion and was ranked among the five largest builders in the United States by market value. Like all large corporations, NVR has its own way of doing things and may not be open to change. In the successful neotraditional projects I've seen, such as Seaside and Newpoint, the developers dealt with small local builders from a position of strength. In Lakelands, where a national builder—Ryan—was involved, the developer laid down very strict guidelines ahead of time. But that is exactly what Jason has not done. Arcadia may have hit the financial jackpot, but at what cost?

21

Mike and Mike

During the early 1900s, Americans could buy so-called mail-order houses from catalogs. The most famous were Sears, Roebuck and Montgomery Ward, but there were dozens of competing companies, and it has been estimated that between 1900 and 1940 more than a quarter of a million houses were built this way.

Rick King, a twenty-five-year NVR veteran, in charge of land acquisition at the regional office, negotiated the New Daleville deal with Arcadia. "We're familiar with southern Chester County, and we think it's a good location," he says. "We also like the neotraditional concept, which allows us to differentiate our product from others. We've had good experiences over the last five years in several neotraditional projects in Maryland and Delaware." King explains that there is another reason his company is interested in neotraditional development. "In most places restrictive zoning is pushing development farther and farther out into distant rural areas, which is less efficient for us in terms of construction. We see neotraditional communities as a useful option for projects in small towns and existing suburbs that are closer to population centers."

NVR is always on the lookout for permitted lots. "We're not in the development business," says King emphatically. "We know how to build, that's our specialty. We're very efficient, and we turn our assets five or six times a year, so we don't want to have money tied up in land." The publicly held NVR builds in eleven states along the Atlantic seaboard, from South Carolina to upstate New York. In 2004 it had orders for thirteen thousand homes—single-family, town homes, and condominiums—ranging in price from $90,000 to $1.7 million. The company doesn't necessarily build only large projects. "An ideal size for us would be about seventy houses, selling thirty houses a year," King tells me. "This justifies the cost of the model homes, and the supervisory and sales staff. But we can do smaller projects if we are already working in an area."

King has one reservation about New Daleville. "It's much more rural than I expected." Arcadia has taken it for granted that the village layout will attract buyers, but both King and Haber single out the exurban location as a potential liability. After all, when people move far out into the country, they expect to live on large lots. Buyers will accept smaller lots, even pay more for them, if there are nearby shops and amenities, but that will not be the case with New Daleville. "With only one hundred and twenty-five houses, there's not enough people to have much of a center," King says.

He admits that NVR paid a high price for the lots. "Arcadia pushed us pretty hard," he says. "We might not have gone along if land in southern Chester County wasn't in such short supply. There are currently eight hundred permitted lots in the county held up by a sewage treatment problem. But we can't wait, we need to keep our machine going. Our houses at New Daleville will start at three hundred and forty thousand dollars and go into the low four hundreds. These are high prices, but Toll is selling *town homes* in this area in the high three hundreds, so we're pretty optimistic."

When Joe Duckworth originally envisioned New Daleville, he expected it would attract first-time buyers. Now the house prices have doubled. I ask King who he thinks the buyers will be. They will be first-time buyers with dual incomes, he tells me, and also "move-ups," that is, people who have just sold a town home or condominium in the booming real estate market and are looking for something larger, newer, fancier. And what about first-time buyers who have not accumulated sufficient equity? They will have to be content with a town home or a condominium apartment, he says, or they might have to settle for an older house—a so-called fixer-upper. Alternatively, they could choose to move to an even more distant exurb, trading a longer commute for a lower house price.

Mike DiGeronimo's main job as town architect is to review the builders' house designs with the aim of assuring that the exteriors contribute to the overall experience of a neotraditional community. This is his first time working with a national builder. "The good news is that NV and Ryan have built in TNDs before, so they know what we're looking for," he says. "The bad news is that it's not easy to convince them to alter their designs. They're open to modifying certain exterior details, but they won't change anything that affects the basic framing." The houses that NV and Ryan have submitted to DiGeronimo have not been designed specially for New Daleville; they are stock "neotraditional" models from NVR's architectural division. What makes them neotraditional is their rather simple, old-fashioned appearance, and the fact that they are narrow enough to fit on small lots. The construction drawings—there are more than sixty sheets per model—are highly detailed and have been engineered to minimize production costs. Understandably, the builders want to make as few changes as possible.

DiGeronimo reviews the designs of Ryan and NV separately. Ryan, founded in Pittsburgh in 1948, is NVR's workhorse brand; NV Homes, which builds chiefly in the Washington, D.C., Baltimore, and Philadelphia metropolitan areas, produces slightly larger and more expensive houses aimed at more affluent buyers. The NV houses tend to be a little more ornate, but the designs of both companies are what DiGeronimo calls generic neotraditional: two-story boxes with pitched roofs and gables, as well as dormers, porches, bay windows, and vaguely Colonial details. Compared with the designs of other builders, NVR's neotraditional models are actually pretty restrained and recall sedate, five-bay, center-hall Colonials—no ersatz Palladian windows, no *Gone With the Wind* porticoes. The less-expensive models, because they are simpler, are more attractive to my architect's eye. The fancier models have front doors with elaborate transoms and sidelights, ornamental gables, and windows topped by keystones and arched headpieces.

Most of DiGeronimo's suggestions are aimed at simplifying the designs. He recommends changing the scale and proportions of the entrances, getting rid of some of the ornamental trim on the façades, and eliminating the decorative molding at the eaves. Ryan generally favors dentil moldings consisting of repetitive little blocks, which DiGeronimo thinks are needlessly complicated. He points out that the size of the shutters is inappropriate—even if they are not operable (the New Daleville guidelines do not require real shutters), they should at least look as if they could cover the windows. He suggests small changes to make the porches look more authentic. Several of the models have two-story balconies but no access from the second floor. DiGeronimo doesn't like this, but there is not much he can do about it.

These are definitely one-sided houses. NVR's designers have lavished attention on the fronts and more or less ignored the sides and backs. Only the fronts have shutters, ornamental details and, in certain cases, brick façades. The arrangement of the windows is orderly and attractive on the façade, but around the corner it's chaotic, with windows, bays, and fireplaces scattered without rhyme or reason. DiGeronimo is concerned about this in the case of the corner lots, where the side is exposed to the street. He makes some sketches that show how windows could be rearranged, cornices carried around from the front, and porches wrapped around corners, so that the houses present a more unified appearance. The sketches are nice, but will the houses turn out this way?

Mike Linthicum, a tanned, fit-looking man in his mid-forties, is in charge of Ryan's Philadelphia West division, which includes Chester County. "We build about three hundred and fifty homes a year," he tells me. "We currently have nine projects under way and twenty in the pipeline." I ask him about the New Daleville design guidelines. "Ryan has built in four TND developments, so we appreciate Mike DiGeronimo's concerns. The two big issues for us are the cost of design changes and how they will affect our production process, since we're a highly systematized company."

Linthicum has been with Ryan for twenty-two years. He explains the company philosophy, quoting Dwight Schar, NVR's CEO. "Our business is like a hamburger stand. We make hamburgers and cheeseburgers. That's it. We don't customize the designs of our houses; except for the specific options that we offer with each model, they're absolutely fixed." Rigid standardization explains why the shutters on Ryan houses are often not sized to the windows. "We use the same shutter for all our houses," Linthicum says. "LRK wants us to use different shutters for different windows. We would have to delete the shutters from the factory order and buy them separately. That will cost more, but since it's in Londonderry's ordinance, we may have to do it."

Linthicum is very conscious of construction costs. The foundation of NVR's corporate success—earnings per share have steadily risen over the last five years—is a highly rationalized building process. Linthicum tells me that Ryan can deliver a house for a construction cost of thirty-six to forty dollars per square foot.* This is an impressive number—most regional production builders manage sixty to seventy dollars per square foot. Large production builders are extremely efficient, with the result that, over the last several decades, although wages have risen and quality has improved, real per-square-foot construction costs have remained remarkably stable.[1]

When I mention Ryan's construction figure to Vince Graham, the TND developer, he says, "That's amazing. I didn't think you could build mobile homes for that price." He estimates that the average construction cost of custom houses at I'On is $165 per square foot. "Buyers are demanding higher-end finishes these days," he says. As well they should, since a four-bedroom house in I'On is currently selling for a million dollars. A comparably sized Ryan house at New Daleville will be a third that price.

Like many national builders, NVR prefabricates the framing for its houses. The company operates production facilities in six states, staffed by their own workforce. These factories manufacture walls and partitions in the form of panels (insulation, exterior siding, and interior plaster wallboard are added on the site). Since the wall panels are not modular, they allow for a variety of designs, and the finished houses do not look like "prefabs." Prefabrication is not primarily an economic measure—panelized houses are not cheaper—but a way to control quality and overcome the difficulty of getting skilled site labor. The house is shipped from the plant to the building site as a "package" of wall panels, roof trusses, plywood, and precut lumber for floors. Later packages include doors and windows, precise quantities of insulation, siding, and so on. "Trim" packages include wood trim and cabinetwork. On-site subcontractors, supervised by NVR

*This does not include site work, such as grading and landscaping, and "soft" costs, such as marketing, development overheads, and profit.

managers, put the pieces together. This process allows NVR to buy materials in bulk, tightly control quality, and reduce theft and wastage.

The idea of precut houses is not new. During the early 1900s, Americans could buy so-called mail-order houses from catalogs. The kits, consisting of precut lumber and trim, cabinetry, even nails, were shipped by rail; local builders did the construction. Although in a few cases entire factory towns were built using mail-order houses, the main buyers of kits were individuals.[2] The most famous mail-order house catalogs were those of Sears, Roebuck and Montgomery Ward, but there were dozens of competing companies, and it has been estimated that between 1900 and 1940 more than a quarter of a million houses were built this way.[3]

Linthicum explains that his company follows a very tight building schedule. "You've heard of the just-in-time principle? Once an order is sent in from the sales office, our plant takes about a week to put together the framing package. It arrives on the site, and the subs have a week to assemble the frame. Then a second package arrives with windows and exterior trim. Once the Sheetrock contractor finishes the inside, the interior trim package is delivered." This intensive process reduces construction time by as much as two-thirds (it takes Ryan only three to four months to build a house). But such a highly organized process also doesn't leave much room for aesthetic adjustments — or, indeed, any other kind. Mike DiGeronimo has his work cut out for him.

Ranchers, Picture Windows, and Morning Rooms

One doesn't buy a house just because of its curb appeal, but the view from the curb is important. It's what we see every time we return home.

The history of the American housing industry is littered with failures. Even the redoubtable Thomas Edison stumbled when it came to housing. In 1908 he patented an elaborate method for mass-producing inexpensive concrete houses. His idea was to erect cast-iron forms for the entire house—walls, floors, and roof—and fill them with concrete through a sort of funnel at the top, all in one pour. After several trials, he abandoned the project.[1] In the thirties, there were a number of attempts to market all-steel houses, though none succeeded.[2] During the forties, the General Panel Corporation invested $6 million in developing and marketing a prefabricated wooden house designed by Walter Gropius, the famous architect and Bauhaus founder. Although the company planned to manufacture thirty thousand houses a year, it produced fewer than two hundred and gave up.[3] In the seventies, the federal government launched the grandly named Operation Break-through, whose goal was to harness the know-how of large corporations such as General Electric, Du Pont, and Inland Steel to produce industrialized housing. It turned out that building affordable and attractive houses was a lot more complicated than putting a man on the moon.

The construction of a typical house requires thousands of board feet of lumber, hundreds of sheets of plywood, and many cubic yards of concrete, square feet of exterior siding and wallboard, bundles of insulation, gallons of paint, reels of wiring, and lengths of copper piping. A builder not only coordinates a dozen different trades to put all this together in a timely and efficient manner but must put it together in a way that is attractive—and affordable—to buyers. He must also judge to what extent buyers are open to innovation. Houses are the largest investments that most families will ever make, and as prudent small investors, they tend to be conservative and to avoid unnecessary risk. While architectural critics frequently disparage the uniformity of housing, that is precisely what buyers demand; they don't want to be stuck with an odd or dated house at the time of resale. Contrarians don't do well in the housing market.

But houses are not only investments, they are homes, and hence sources of personal pleasure and pride. Like clothes, they convey status and social standing; like cars, they tell people something about their owners. Thus, the decision to buy a house is emotional as well as financial. I understood this the first time my wife and I went house hunting. There were houses that we knew, from the second we saw them, were not for us; we didn't need to go inside. Of course, one doesn't buy a house just because of its curb appeal, but the view from the curb is important. It's what we see every time we return home.

The schizophrenic house buyer is both a status seeker and an investor. In addition, he or she is a consumer. Renovating a kitchen, for example, is done with one eye on convenience and one eye on resale, as well as a glance at the attractive advertisements in the latest issue of *House & Garden.* The demand for consumer goods being driven by what's new and up-to-date—the latest thing—the house buyer is not immune to fashion.

Although the desire for novelty is generally tempered by an inclination to the safe bet, there was one period when buyers let their hair down. Buoyed by the post–Second World War boom, optimistic about the future, and gripped by the idea of Progress, Americans embraced innovation as never before, in the way they traveled, the way they brought up their children, in their manners—and in their homes. The hallmark of that period was the ranch house. It is said to have been invented in 1932 by Cliff May, a self-taught San Diego architect, but it also owed a debt to Frank Lloyd Wright's Usonian houses, and to Alfred Levitt's "Levittowner." Today the suburban ranch house is considered the epitome of conservative taste, but at the time it represented a radical departure from tradition. To begin with, all the rooms were on one floor. The layout was open and casual, with wood paneling instead of wallpaper, and room dividers instead of interior walls. The exterior was unabashedly contemporary and did away with steep roofs, dormer windows, and porches.

The casual, spread-out ranch house (it was also known as the California ranch and the rambler) had enormous appeal and by 1950 accounted for nine out of ten new houses.[4] In hindsight, the rancher's most striking feature was its diffidence. Low to the ground, it lacked traditional domestic status symbols, such as porticoes and tall gables. Its one extravagance was a large window facing the street—the picture window. As far as I have been able to determine, picture windows made their first appearance in Levittown, Pennsylvania. Alfred Levitt had already used floor-to-ceiling walls of Thermopane glass to open the house up to the backyard, but in the "Levittowner" he put an eight-foot-square kitchen window facing the street. The common criticism that picture windows offer neither privacy nor a view misses the point. Picture windows are meant not for looking out but for looking in.[5] They are displays—for Christmas ornaments, Halloween skeletons, and Thanksgiving Day wreaths. Opening the house up to the street—something that neither Wright nor modernist architects did—is a curiously disarming gesture. "Feel free to look in," the picture window announces, "we have nothing to hide."

Another design innovation of the fifties was the split-level house, which mated a ranch house with a two-story section, half a flight up and half a flight down. The split-level originated in California as a way of building on slopes, but it also provided useful solutions to two new domestic problems. One was where to put the television. The first televisions, which were designed like pieces of furniture, stood in the living room. As television watching became increasingly popular—especially among children—to preserve the living room for formal entertaining, the set was moved to its own special room: the recreation, or rec room. The rec room was usually in the basement, but in a split level, this was only half a flight down, less drastically separated from the rest of the house. The other problem was where to put the car. Using the lower floor of a split level as a garage was an inexpensive alternative to the attached carport. A variation of the split level was the bi-level, which had a half basement (for the bedrooms) and located the entrance halfway between the two floors. By 1970 four out of five new houses were either ranchers or splits.[6]

When I was growing up in Canada, my friends and I lived in new, ranch-type houses. Ours was not large. It contained a living room, an eat-in kitchen, three bedrooms, and a bathroom, all in less than eight hundred square feet. This would have been a tight fit for the four of us, except that we also had a large basement, which accommodated my train set, my parents' collection of *National Geographic*s, and a laundry area. There was also a rec room—my first design project. I nailed sheets of textured plywood to the walls, laid vinyl tile on the floor, and stapled perforated acoustical tile to the ceiling. I built a small bar in the corner and a Mondrianesque room divider. I was eighteen, and I thought it the height of chic.

Housing has always been governed by a simple rule: as people become richer, they spend more money on their homes. Historically, this has meant using more expensive materials—varnished mahogany instead of painted pine, marble instead of brick—updating the décor, or adding technological refinements, such as gas lighting, indoor plumbing, or central heating. In addition, spending more money has usually meant making the home bigger. This happened in Renaissance Italy, seventeenth-century Holland, and nineteenth-century England. It also happened in the prosperous second half of the twentieth century in the United States.* Some statistics: in 1950 the median size of a new house was 800 square feet; by 1970 this had increased to 1,300; twenty years later it had grown to 1,900; and in 2003 it stood at 2,100. More than a third of new houses built today exceed 2,400 square feet.[7]

My childhood bedroom was about 8 feet by 10, just big enough for a bed, a desk, and a chest of drawers. My younger brother's room was smaller. The most generous space in our home was the garden, where my father grew gooseberries and crab apples. Our 60-by-100-foot lot was small for the time. My best friend, whose father was a local grandee—the manager of Woolworth's—lived nearby in a larger ranch house, which occupied a much bigger lot. That's the thing with ranchers; as they get larger, they stretch out and need more space. A 2,000-square-foot ranch house with a two-car garage, for example, needs a lot at least 120 feet wide.

By the nineteen eighties, buyers wanted larger houses, but Proposition 13, which required developers to pay for their own infrastructure, had made land much more expensive. The builders' solution was to return to two-story houses, which don't need such large lots, and which are up to 30 percent cheaper to build because of smaller foundations and roofs. Today, more than half of all new

---

*In 1950 the median national household income was $3,000, or about $20,000 in modern dollars; today the national median income for married-couple families (who are the majority of home buyers) is $60,000.

houses have two stories.[8] But that is only one change. No one builds ranch houses or split-levels anymore. Picture windows and carports are gone, so are breezeways. Home buyers' affair with modernistic design is over. Ryan Homes has a telephone-directory-size catalog of all its models currently in production, more than 150 of them. When I leaf through the pages, I notice that all the houses have similar architectural features: pitched roofs, gables, dormers, bay windows, keystones, shutters, porches, and paneled doors. Americans' fondness for such conventional imagery is characterized by some critics as nostalgic and retrograde. In fact, it represents a long domestic tradition that extends to colonial New England and Virginia. In that history, the brief fling with the rancher was an anomaly.

The practice of giving names to house models, which originated with the mail-order house manufacturers, has continued to this day. For example, Ryan names its neotraditional models after famous writers: Hemingway, Faulkner, and Fitzgerald. Linthicum has chosen four models for New Daleville: the (Sidney) Sheldon, the Melville, the Michener, and the Carroll. I don't say anything to him about this distinctly odd literary round table. Which models to build and how much to charge for them are crucial decisions. "The Sheldon is the smallest and will be our loss leader," he says. "It'll allow us to advertise a low price, in the upper three forties. The Melville is slightly larger, a sort of center-hall Colonial that will sell in the mid–three fifties. The Michener is next, and the Carroll will be our largest and most expensive product, in the low four tens. We'll build a furnished model of the Carroll. Not everyone will be able to afford it, but you need something that will make an impact so that people will remember you."

Since Ryan is already advertising New Daleville on its website, I take a look at floor plans of the Sheldon. At 1,944 square feet on two floors, it's no McMansion and is slightly smaller than the national median for new houses. Like all the models at New Daleville, it comes with a two-car garage, detached or attached, depending on the type of lot. The first-floor entry foyer is flanked by a living room and a dining room. This is a traditional arrangement (similar to that of my own house), but what is distinctly untraditional is the living room's size—it's one of the smallest rooms in the house. In the Sheldon, the buyer can even dispense with the living room altogether by adding optional French doors and turning it into a study. Starting in the eighties, the living room, which had been dominant since the beginning of the century, has been usurped by the family room. A successor to the fifties rec room, though no longer in the basement, the family room is a casual living space next to—and usually overlooked by—the kitchen. At first, the family room was merely a convenient place to park the television, but as formal entertaining fell out of favor and informality prevailed, people found themselves spending more and more time in their family rooms, which grew accordingly. The proximity to the kitchen had the added advantage that whoever was cooking no longer felt isolated. The Sheldon's family room is the largest room in the house, about thirteen by sixteen. A telling sign of the shift in the domestic center of gravity is the location of the fireplace, the domestic hearth. In the Sheldon, as in most Ryan models, the fireplace option is available only in the family room.

Options, which give buyers the opportunity to personalize their homes, are an important part of all production builders' marketing. "Extension options" add space to the house. In the Sheldon, the family room and the kitchen can be extended by two feet. Or a "morning room," a sort of breakfast room, can be added to expand the kitchen. Interior trim options include "tray" and "cathedral" ceilings, which raise the height of the room, crown moldings and chair rails, finished basements, and upgraded kitchen cabinets. It is also possible to upgrade the kitchen appliances. (Central air-conditioning, by contrast, is a standard feature.)

There are many window options: bay windows in the living and dining rooms, extra windows in the family room and kitchen, and additional windows and skylights that turn the morning room into a full-fledged sunroom. Ryan may sell only hamburgers and cheeseburgers, but it offers a choice of ketchup, mustard, or relish, according to taste.

I ask Ryan's regional sales manager, Carmela Bond, about the most popular options. "The common choices are those things that are difficult to upgrade later, like flooring or bathroom finishes. Almost everyone upgrades their kitchen cabinets, and fireplaces are very popular," she tells me. The cost of these options varies. An ordinary window adds $450, but a bay window is $3,000; kitchen cabinet upgrades are $2,000 or $3,000; a front porch can be more than $8,000; a morning room costs about $11,000, while a finished basement adds almost $15,000. According to Bond, buyers routinely add $30,000 worth of options, and in some cases this figure will be twice as high, though rarely exceeding $100,000.

The Sheldon comes with either three or four bedrooms on the second floor. What is unusual, compared with production houses of twenty years ago, is the size of the master bedroom suite. The twelve-by-twenty bedroom is connected to a walk-in closet and a bathroom, which comes in different versions, the largest with two sinks, a Jacuzzi bath, and a shower stall. The smaller bedrooms share a hall bathroom; there is a powder room on the first floor and an optional powder room in the basement. This abundance of bathrooms is typical; half of all new homes built today have two and a half bathrooms or more.[9]

Once the buyers of a Sheldon have picked a floor plan, they must choose among three façade options, all more or less Colonial and all, to my eye, simple and attractive: a traditional five-bay window arrangement with a central front door; a variation that adds a central gable and a roofed entrance; and a third version with a porch across the entire façade. The wall material is either vinyl siding or, at considerable extra cost, brick. The only other façade option is to dress up the roof with false dormers.

According to NVR's literature, "Our homes combine traditional or colonial exterior designs with contemporary interior designs and amenities."[10] The Sheldon is just such a blend of tradition and modernity. The façade is resolutely old-fashioned—it would not look out of place in Colonial Williamsburg. But while the house appears to be a center-hall Colonial, the interior is actually open, offering an unobstructed view from the front door to the back of the house, and an unbroken connection between the family room and the kitchen. This openness recalls the interior of a Wright Usonian, or a Levittowner. After the adventurous fifties and sixties, it appears that American home buyers have settled down to have their cake and eat it, too.

There's nothing about the hybrid design of the Sheldon that is particularly unusual. The chairman of the board and CEO of NVR, Dwight C. Schar, once told me, "We don't lead parades, we follow them." Schar is skeptical of novelty and mentioned skylights, trash compactors, and central vacuum systems as examples of fads that came and went. "It takes twenty years before a design innovation becomes part of the mainstream. Look how long it took for town homes to be popular." What about living rooms? "If people can afford it, they want one," he said. "Even though the living rooms in eighty percent of our houses have little furniture. Our buyers also want a formal dining room, even if they mainly eat in the dinette." NVR, like all builders, can't afford to misinterpret or ignore home buyers' priorities, be they desires for a family room, a fancier kitchen, or walk-in closets. In that sense, at least, the house is a quintessential consumer product—the buyer is always right.

Pushing Dirt

Improving raw land is a messy and unpredictable business.

Since the planning commission has not reviewed the builders' houses, the first glimpse Tim Cassidy has of the designs is when he accidentally comes across a set of architectural plans in the township office. He immediately e-mails Mike DiGeronimo. "Just at a glance, there are some details that I am sure, in LRK parlance, fall into the category of 'Don't Do This.' Also, side elevations were supposed to be designed such that when a material, such as brick or stucco, is used on the front façade, it is carried around the corner to a transitional element, such as a chimney or a bump-out, on the side elevation. Again, a quick glance indicates that the material changes occur directly at the corner of the façade. I am going to review the drawings this weekend, but I was wondering whether or not LRK had any comments."

"I sent Mike the e-mail to goose him and remind him that that we're still here," Cassidy tells me. "I didn't want him and Jason to think that they were done with us." Architectural design is on Cassidy's mind since he has recently left Tom Comitta's office to join a local architectural firm. "I've been with Tom, on and off, for eighteen years. I started as a newly minted landscape architect, then decided to study architecture. When I came back from Texas,

I got married, and started a family. So I resumed working for Tom. But I always wanted to do architecture."

It turns out that the drawings Cassidy saw were stock plans that didn't incorporate any of the exterior changes being discussed with DiGeronimo. Cassidy is mollified but wary. "Mike claims that he's working with the builders and that 'revisions are pending.' Who knows? I've asked the township not to issue any building permits unless there is a letter from LRK stating that the plans are in compliance with all the design guidelines. So, we'll see where that goes. The township may not have direct control over design, but there are some permitting matters outstanding, so we still have some leverage." I tell him I'm pleased to hear that he's still involved in the New Daleville project. "Of course I'm involved," he answers. "I live here!"

Mike DiGeronimo has almost finished reviewing the house designs. He's had some small victories. The shutters will be properly sized. In response to his suggestion, the Ryan staff has agreed to modify the side-facing corners. They will not redesign the elevations, but they will make options such as wraparound porches and bay windows mandatory on corner houses. They also agree that gas fireplaces on walls exposed to the street will have stucco chimneys, even though these will not be functional. DiGeronimo is not opposed to one-sided houses, but he doesn't like abrupt changes of details and material at the corners, and Ryan goes along with his suggestion to return cornices on the sides and to carry brick eight feet around the corner. "This will raise the house price, so we'll probably limit the number of brick options," says Ryan's Mike Linthicum. "The truth is that brick is a lot more expensive than vinyl, adding more than fifteen thousand dollars, so not many people choose it. Our customers prefer to spend extra money on interior upgrades. But since we do want some variety, we'll probably price the brick option low to encourage a few buyers."

On the whole, however, the builders have resisted major

changes. Their houses are designed to optimize NVR's highly rationalized production system, and they incorporate many small refinements, the results of feedback from their building supervisors, sales staff, and customers. Unlike the smaller production builders that DiGeronimo usually deals with, the NVR companies not only are much larger but have a lot of experience building and selling houses in TNDs, and they've learned what their buyers are willing to pay extra for. While Linthicum appreciates that architectural guidelines are part of the process, he's not willing to make any changes that will jeopardize sales. Although he doesn't say this, I suspect that the high price Ryan paid for the lots means he is under pressure to make sure the houses sell quickly. He refuses to change the design of the entrances or to get rid of the dentil molding at the eaves, for example. The arched windows and the keystones that DiGeronimo doesn't like are staying, too. "Deleting them would mean explaining to buyers why their houses are different from the brochures, since we're not going to print new ones," Linthicum says. "Anyway, I actually think the houses look better with these features. If we take them out, the façades will be too plain for our customers."

The simplest versions of the Sheldon and the Melville are attractive in a countrified sort of way, especially when equipped with porches. But more expensive models, such as the Michener and the Carroll, particularly with brick façades, have curved broken pediments over the front doors, elaborate transoms, and brick quoins that seem more suited to Rittenhouse Square than to rural Chester County. When I ask DiGeronimo about this, he says they will probably look okay. He sounds resigned.

The architectural handbook that Arcadia and LRK created to reassure the township and Tim Cassidy about the quality of design of New Daleville claimed the houses would be inspired by the villages of Chester County. Linthicum tells me that in some projects he has been able to convince the head office to develop new models, but since doing this costs about $40,000, it's hard to justify for a project as small as New Daleville. The truth is that, despite DiGeronimo's efforts, the New Daleville houses will be similar to the neotraditional models NVR builds from South Carolina's Lowcountry to upstate New York.

"New Daleville is a remarkably picturesque neighborhood set in the rolling hills," reads the Ryan Homes website. "Reminiscent of old-time neighborhoods, this lovely neo-traditional community has a central boulevard lined with picket fences leading you into tree-lined streets and alley ways. The lush landscape is laced with bench-lined paths and winding walkways to pocket parks and recreation areas where families and friends can gather and have fun." On this sunny day in mid-April, New Daleville doesn't look anything like that. The earthmoving crews have been working on the site for a week, and what was once a picture-postcard landscape of fields and pastures is now a sea of red earth. Jim Weidner, an athletic-looking man whose hobby is long-distance bicycle racing, is Arcadia's construction manager. He takes me around and explains the work involved in preparing the site for the home builders. The first task is to deal with surface runoff, that is, rainwater, since environmental regulations require that all rainwater must be kept on the site. The water will be collected from the streets by a system of underground concrete pipes and drained into a retention basin, where it will sit long enough for the silt to settle. It will then pass through gravel filters before being discharged into the existing wetland area. The aim is to keep silt from clogging the wetlands.

There are two earthmovers and five assorted bulldozers, backhoes, graders, and rollers scooting over the raw earth. This contractor builds interstate highways as well as suburban office parks and regional schools, so rolling out fifty acres of roads and sewers must be a piece of cake. The retention basin at the low end of the site is contained by earthen dams called berms. Temporary black-plastic diversion fences slow down runoff until the basins are operable. Orange security fences keep the earthmoving equipment away from the wetlands and the future drip field. "Once the basins are finished, they'll bring on a second crew to build the streets," Weidner tells me. "Then it'll get really busy." From a distance the machines looked like toys in a sandlot; close up they are enormous. A noisy grader rumbles past. Weidner points to saucer-like hardware on the roof. "That's a satellite dish connected to a Global Positioning System. Inside the cab, a computer programmed with the topographical plan of New Daleville gives the operator a precise readout of the terrain, so he knows how much to cut or fill."

We walk along the flat top of a berm that resembles an ancient bulwark. Once the basins are finished, the berms will be seeded and the crews will turn their attention to grading and compacting the streets and creating what Weidner calls the "pads" for the individual house lots. Storm water and sewer pipes will be buried, along with other utilities (houses at New Daleville will have individual underground propane tanks). Then, the streets and lanes will be paved and the granite curbs installed. Sidewalks will be built as houses are completed. "Our deadline is September 1," he says. "If we miss that date, the home builders can theoretically walk away. We've had a slow start because of rain, which meant that the berms wouldn't compact properly. In another project, we had so much water in the soil that we had to drill wells to pump it out. Fortunately, that wasn't the case here." Weidner expects to have the pads ready in June or July, when the contractor will begin grading the streets. "We'll start with the boulevard, which is where Ryan and NV will build their model homes. Then we'll do the side streets and the back lanes," he says.

The topsoil has been scraped off and stored on the future township park, which now resembles an earthen mesa. Since most of the site had no trees, the churned-up landscape looks especially desolate, as if someone had dropped a bomb onto Dr. Wrigley's pretty cornfield. "It's scared my neighbors to death," Tim Cassidy later tells me. "Their unspoken message is 'How could we have approved a plan like this?' Oh well, I'm sure they'll calm down once the grass starts to grow." But I sense that even he is shaken by the extent of the disruption.

Three months later the grass is growing on the berms, the storm water and sewer pipes are buried, and the roads are graded. The central boulevard, which is the main feature of the plan, is clearly discernible, its wide median planted with grass. Concrete street drains poke out of the earth, and the future streets are churned into mud by the massive tires of the earthmoving equipment. The finished house pads are anything but flat; they are actually several feet lower than their final elevation, since they're designed to accommodate the earth that is excavated from the basements and foundations. I was expecting something neater. But Weidner sounds pleased and tells me that the builders are very happy with the way the grading has been done.

Work has not yet started on the drip irrigation field. The state environmental agency has already given its planning approval, but because New Daleville is discharging treated wastewater into the ground, it also needs a federal permit. This permit consists of two parts: the first certifies the quality of the effluent; the second is basically a building permit for the treatment plant. Arcadia has the first but not the second. Weidner says that the reason for the delay is unclear but that he's not worried. It will take twelve months to build the plant and the drip field, which means it won't be in operation when the first houses start going up. This sounds like a complication, but Weidner explains that the treatment plant cannot function properly at less than about 50 percent capacity, so until half the houses are occupied, all sewage will have to be trucked away from the site.

It's now June 2005, but water supply continues to be an issue. Two and a half years have passed since Arcadia first talked to Chester Water, and construction of the trunk line has still not started. This has nothing to do with the opposition of the neighboring township, which, as the lawyers predicted, has faded away. The difficulty concerns a delay in the negotiations between Chester Water and the Honeycroft Village developer, who seems to be in no hurry to sign. Since Honeycroft is the lead client for the trunk line, Arcadia can't finalize its own contract, and without a contract it can't apply for a permit to build water mains. Since the contractors have finished grading the site, they are ready to lay pipe—or leave. To avoid further delays, Arcadia decides to do something a little risky: install the water mains without a permit. "We think the risk is minor," says Jason Duckworth. "The problem with Honeycroft is simply one of timing; they are definitely going to build the pipeline. We have the township inspector on site, and we're videotaping our installation so that Chester Water can give us a retroactive permit." He is upbeat, as usual, but the delay is potentially serious. The construction of the Chester Water pipeline is expected to take two to three months, and the longer it is postponed, the later the water will arrive. And without water, homes cannot be occupied. "We can pump sewage out," Weidner says, "but we can't truck water in."

When New Daleville was in its early planning stage, three years ago, I remember Joe Duckworth showing me one of the worksheets that outlined the schedule for the project. The chart resembled a staircase, each step neatly following the previous one: rezoning, township approval, state permits, land acquisition, site preparation, model homes, and so on. The actual process hasn't been like that at all. The sequence has gotten jumbled—it's now like a twisted stair in an Escher drawing, seeming to go up but frequently going sideways or down.

Yet there is progress. In July, the builders bring in two double-wide trailers that will serve as temporary sales offices until the model homes are built. The trailers are set up in the space that will be occupied by the commercial building. NVR erects a gazebo between the two offices and puts up a white-painted horse fence around the parking lot.

Jason doesn't like the fence. "Our landscape architect has designed several structures that include a gazebo and two pergolas. We plan to build these out of unpainted barn wood—we've actually bought an old barn. The idea is to make New Daleville look a little more rustic. We might use barn wood for the street signs, for example. We're also thinking of buying an old windmill tower and installing it as a landmark. We suggested that the builders also use cedar fences to fit in with our theme, but they didn't want to. They also rejected our suggestion to use unpainted wooden fences around the houses. They're concerned about maintenance, and their corporate policy for neotraditional developments is to use white vinyl picket fences. They're going to build fences along the length of the boulevard. I'm not happy with it, but they're paying for them. At this point, there are some things that are not worth arguing about."

There has been friction between Arcadia and the builders on other issues. "For example, we can't agree on the 'weekend directionals,' the roadside signs that will direct buyers to the sales trailer," says Jason. "We want to emphasize New Daleville, but they have a standard template that features their own brand. Similarly, in the sales brochure, we want to highlight community issues but they want to focus on the houses. What it really comes down to is: Whose development is this, anyway? They've bought the lots, so they feel it's now their operation. But we still have a vested interest in the overall concept."

It's understandable that NV and Ryan now think of New Daleville as *their* project. They've paid a lot of money for the lots, and they believe that they know best how to market the development. As the risks and rewards shift from the developer to the home builders, the latter want to be in control. Judging from the way NV and Ryan have handled the design guidelines, they will probably get their way.

Although the builders have set up sales trailers and are advertising New Daleville in local newspapers, they haven't actually bought any lots from Arcadia. "Without a water contract, we can't get a permit for the water mains, and without a permit, the builders won't buy any lots and start building their model homes," says Jason. "Everything is held up."

It's now September. Honeycroft and Chester Water have finally reached an agreement, but Arcadia still doesn't have a contract with the water company. "They're unhappy with the fact that we went ahead and put in the water pipes without their approval, so now they're dragging their feet," says Jason. "They're going to send an inspector to review our videos of the installation. He'll probably want to open up a few holes to examine the work. Although we've put in the street curbs, we've held off paving the streets, just in case. If the contractor did his job right, there won't be any trouble." Meanwhile, the weather is turning colder. The only good news is that DEP has issued a building permit for the treatment plant, and the township has finished reviewing the builders' house plans and has issued building permits for the model homes. Building permits in hand, and assured that the water permit is imminent, the builders start work on three model homes.

It is January 2006 before New Daleville gets its water agreement. Chester Water insists on inspecting the individual connections between the houses and the street mains. At one point it appears that the water company may demand to check each of the 125 house connections, but they finally settle on a sample of half a dozen. After several more delays, the inspection is completed without a hitch. Since it is now midwinter, paving of the streets cannot resume for at least another three months. Weidner figures that the testing delay alone has cost Arcadia $20,000. But at least the trunk line that will supply water to New Daleville is finally under construction.

24

The Market Rules

House prices generally rise slowly, reflecting rising construc-
tion costs, increased demand (due to population growth and
obsolescence), and inflation. During a housing bubble, by
contrast, the increase is significantly greater as people bid up
the prices of houses on the assumption that they will be worth
considerably more in the near future. In other words, they
act more like speculators than like home buyers.

For the last decade, the United States has been experiencing a housing boom. In 2004 sales of single-family houses passed 2 million, a record. But not only are more Americans buying houses, they are also willing to pay more for them. In the last four years, new home prices have risen almost 50 percent, which is dramatically more than the historical average of about 5 percent a year. These are national figures; the four-year increase in selected markets has been even more dramatic: 92 percent in Miami, 103 percent in Los Angeles, and 119 percent in San Diego.[1] Such spectacular and unprecedented jumps led some economists in the summer of 2005 to characterize the present housing market ominously as a "bubble."[2]

An economic bubble is defined as "a situation in which excessive public expectations of future increases cause prices to be temporarily elevated."[3] House prices generally rise slowly, reflecting rising construction costs, increased demand (due to population growth and obsolescence), and inflation. During a housing bubble, by contrast, the increase is significantly greater as people bid up the prices of houses on the assumption that they will be worth considerably more in the near future. In other words, they act more like speculators than like home buyers. Since the high prices are not justified by what economists call the fundamentals (population growth, construction costs, inflation), sooner or later they have to stop rising. At that moment, the expectations that have been fueling the market are dashed, and as panicky investors sell, prices fall precipitously and the bubble "bursts."

The media love the idea of a bursting housing bubble, which seems to go along with the rest of the bad news: the faltering war in Iraq, Hurricane Katrina, and a series of business and political scandals. At the same time, the evidence is inconclusive, and many of the newspaper reports are contradictory. Sales are up, they're down; prices are up, they're down; builders are selling more houses, they're selling fewer. I decide to ask an expert. "There is no single-family housing bubble," says my friend Peter Linneman, who is a professor at the Wharton School and also manages a real estate consulting firm. "It's just a bubble of stories about a housing bubble." He agrees that there has been an unusual run-up in prices, but he sees this not as the consequence of a buying frenzy but as the logical result of the market.

Here is his explanation. Over the last four years, mortgage rates have declined 23 percent. Lower mortgage rates mean that more people can afford to buy a house, and Linneman's research shows that for each 1 percent drop in rates, housing demand increases 1 percent. Thus, the current rate decline has produced a 23 percent increase in housing demand. Household disposable income has a similar relationship to housing demand; as incomes rise, more people can afford a house. Since real incomes have risen approximately 12 percent in the last four years, housing demand has gone up a similar amount. Population has increased by about 4 percent

over the same period, which likewise fuels demand. Last, an inflation rate of 8 percent over four years can be expected to raise home
prices a similar amount. Adding these figures yields an overall
increase in demand over four years of 47 percent.

On the supply side, in the same time period, 2 percent of existing houses were destroyed by fire, demolition, or abandonment,
while builders added only about 6 percent. This surprisingly
small increase in the housing stock is caused by local constraints
on development, particularly in California and the Northeast. (In
Chester County, for example, despite consistent demand, fewer
building permits for single-family houses were issued in 2004
than in 2003, and fewer in 2005 than in 2004.) Thus, while national
demand has risen 47 percent in four years, supply has increased
only 4 percent. "Given the massive imbalance between real
demand and supply, it's no wonder that housing prices have risen
dramatically," Linneman says. "Home prices have risen exactly as
one would have expected."

Linneman makes it sound rational, but the housing market is an
odd animal. For example, there are consumer guides for buying
computers, cars, and dishwashers, but none for buying homes (J.
D. Power and Associates rates individual builders, but not specific
home models or projects, for customer satisfaction). Despite the
fact that a house is the largest investment most families ever
make, the process is basically informed by gossip and guess-
work. A paper written by three economists from Pomona College
puts it this way: "The residential real estate market is populated by
amateurs making infrequent transactions on the basis of limited
information and with little or no experience in gauging the fundamental value of the houses they are buying and selling."[4]

Home buyers may be amateurs, but they know something
important about the housing market that seems to have eluded the
media: the sticker price is not necessarily the most important
factor. The cost of ownership is a function of interest rates, taxes,
and the cost of utilities. Low interest rates and higher household

incomes (which allow families to devote a larger percentage of income to housing) have made expensive houses more affordable.[5] The cost of owning is also a function of what economists call "the opportunity cost of capital," that is, the income one would have received if one had chosen to be a tenant rather than an owner. When interest rates are low, mortgages are cheaper and opportunity costs are low. Simply put, buying a larger house makes more sense than leaving one's money in the bank or investing in a lackluster stock market.

If interest rates go up, alternative investments become more attractive. But homeowners do not react to changes in interest rates—or house prices—like stock traders. The so-called transaction costs of selling a house and moving to another—agents' fees, legal fees, taxes, moving expenses—are high. Moreover, housing prices are what economists call "sticky." Sellers are loath to lower prices when demand slows; they stubbornly hang on. They will often take a house off the market and wait for a more propitious opportunity rather than accept less than what they have come to believe their home is worth. In any case, they have to live somewhere. Which is another reason that Linneman doesn't agree that the present housing market is a replay of the 1999 technology bubble. "Homes are fundamentally different than dot-com stocks, since they generate a housing service flow. That is, we live in them. Whereas the absence of cash flow meant that, when the gloss was gone from dot-coms, there was nothing to hold up share prices. Even if the gloss disappears from housing, the service flow of living in one's house will support housing prices over the long term."

Housing bubble or not, in early 2006 there are signs that the gloss is starting at least to tarnish. The current thirty-year mortgage rate is slightly more than 6 percent and rising, compared with slightly less than 6 percent a year ago. Demand for new houses is weakening. Single-family house construction for the month is a couple of percentage points lower than a year ago, while the issuance of building permits for houses, an indicator of future construction activity, is down 5 percent.[6] According to a front-page article in *The New York Times* in October 2005, "The question remains whether all of this represents a momentary cooling off of some overheated housing markets, or it presages a more pronounced downturn that would end a decade-long boom."[7] Toll Brothers has announced that it expects to sell fewer houses over the next year than it had earlier predicted. "It appears that we may be entering a period of more moderate home-price increases, more typical of the past decade than the past two years," Joe Duckworth's old boss, Robert I. Toll, is reported as saying.[8] So, whether or not there is a bubble, New Daleville is seeing the light of day under what may be less than ideal circumstances.

At the Arcadia Land Company, there is no sign of a downturn, quite the opposite. Jason Duckworth has just rented the entire second floor of the old bank building in Wayne to accommodate a staff that now numbers ten people, including an engineer and another Wharton grad. The extra people are needed to manage a growing roster of projects. Woodmont, which is Arcadia's first TND, is well under way. The zoning change for the neotraditional neighborhood in Sadsbury has been slower, despite the bus trip to Kentlands, but recently that project, too, received final approval, and it is moving ahead.* As for Joe Duckworth, he's already immersed in his next deal, and it's a big one. He shows me the plans for a new community called Bryn Eyre. The old industrial site, which once belonged to Bethlehem Steel, covers 3,500 acres, or five and a half square miles. Arcadia is one of several partners, but is taking a leading operational role. The project has the support of the local borough, the county, and the state, and the permitting

---

*Not without incident; during the supervisors' final vote, the township building had to be evacuated because of a bomb threat.

process is going smoothly. Construction is slated to begin in two years. "This will be a real town. We'll have twelve thousand homes, four schools, and two downtown areas," Duckworth tells me excitedly. "It'll be the biggest residential development in Pennsylvania since Levittown."

Duckworth wants to talk about Bryn Eyre, but I steer him back to New Daleville. Does he have any concerns about a possible downturn in the housing market? "I feel solid about the situation," he says. "Job growth is good, and there's still a limited supply of permitted land in Chester County. Even if the mortgage rate goes up to seven percent, that's still a good rate. Of course, if it hits ten percent, then we have a real problem." Duckworth knows what he's talking about. "I went through two periods when interest rates went up," he says. "One was in 1990–91, which wasn't too bad. The other was in 1980–81, which was awful. Mortgage rates went to fourteen percent. In our region, demand dropped by half!" Market corrections occur regularly in the development business. I ask Duckworth how a builder copes with a severe downturn. "If you haven't sold any houses in a project, it's easy, you simply chop prices. But if you're caught in the middle of a project when the market stalls, it's different. You can't just reduce prices, since this really upsets the people who have just bought a house from you. If they see their property values dropping, they come to you and complain, and they bad-mouth the project to potential customers. So what you do is what I call increasing the value proposition for the buyer. You don't change the price, but you offer more. You make previous options standard. If there's a kitchen upgrade, or a bonus room, you make it part of the package. If window dressing isn't enough, you effectively reduce the price by subsidizing the mortgage, or by offering a mortgage-free period, say six months, as an incentive."

What happens if the downturn continues, I ask him. "Then you have to try something else. If you're offering models A, B, and C, you add a model D, which you do price lower. But it's not the same house, so the previous buyers don't mind. The new house can't look crummy, though, since that would be perceived as lowering property values. You can build a smaller house, for example, but the same width as the larger models, so it looks similar from the street. You can also have fewer features inside the house, or less expensive trim. If things get really bad, you can offer incentives, and even negotiate lower prices. The advertised price stays the same, but you sit down with the prospective buyer and basically say, 'What would it take to close this deal?'"

Earlier, when I spoke to Carmela Bond, Ryan's sales manager, she told me that in the three months since the sales trailers opened, the number of visitors to New Daleville has been very small, although she assured me that Ryan considers it a pioneering project and is willing to be patient. Should they be worried, I ask Duckworth. "Well, the first lesson is that it's hard to sell anything out of a trailer, especially a new idea like traditional neighborhood development. You have to get the models up so that people can experience the whole package. I'm convinced that trailers are a waste of time." He says that he and Jason have discussed this, and they now realize they made a mistake in their dealings with the builders. "I assumed that they knew what they were doing, so although I made suggestions, I tended to defer to their judgment. I was wrong. I may not have built any TNDs, and I may know less than the corporate head office, but I've visited a lot of projects around the country, and I do know more than the division heads who are running these projects. They tend to think that TNDs simply require a different kind of ordinary house, and they want the execution to be like a conventional project. Next time, we're going to require builders to complete the model homes *before* they begin selling. You have to show people what the street will be like, and what it means to have compact lots."

Duckworth also thinks that Ryan may have set its house prices too high. "I would have expected the starting price to be below three hundred thousand dollars. Instead they advertised the project as starting in the upper three forties. They recently reduced it to the three tens, but in my opinion that's still too high." He expects sales to remain flat for the moment. "We're now into December, and nobody buys a house in December or early January. But the six weeks between mid-January and the end of February are a hot period. By then, they'll have the models up, and that'll be the real test. I expect them to each sell a house a week."

What happens in the case of a major downturn? "NVR has a relatively small deposit, so they could walk away," he says. "At that point, you don't deny reality. We would probably have to renegotiate the price of the lots. I might offer a lower price, until things pick up, after which they would have to give me a larger share. I'm pretty confident that NVR will negotiate in good faith. I now control thousands of lots at Bryn Eyre; builders are anxious to keep on my good side. Rick King has already talked about taking three hundred lots a year from us. So they'll be patient at New Daleville."

25

Bumps in the Road

In March of 2006, the Financial Times *reports that sales of new homes in the United States have dropped further, and the stock of unsold homes has hit the highest level in ten years. Not the bursting bubble that so many have been predicting, but definitely a slowdown.*

Early in the new year, to encourage sales, Ryan further reduces house prices at New Daleville, lowering the starting price to $290,000. As an additional incentive, buyers who finance their homes through NVR's in-house mortgage company are offered a free finished basement, or the equivalent in other options. Mike Linthicum juggles Ryan's lineup, removing the expensive Carroll and adding two midrange models, called the Fitzgerald and the Austin. "We are monitoring and adjusting as necessary," the sales manager, Carmela Bond, tells me. She sounds hopeful but raps her knuckles on the desktop. Touch wood.

By the end of February, just as the first model homes at New Daleville are being finished, there are more signs that the previously strong national economy is faltering: interest and inflation rates edge up, and consumer expectations dip. The Conference Board, a business group, reports that national sales of new homes have fallen 5 percent in January, the sharpest drop in two years, and that the inventory of unsold units is up a third from the previous year.[1] For the reasons that Joe Duckworth explained, house prices have not fallen dramatically, but it appears a change is in the air. "It's pretty unambiguous that demand for housing has been weakening in the last couple of months," an economist at UBS, a global financial firm, tells *The New York Times*.[2] This glum view seems warranted when, less than a month later, the *Financial Times* reports that sales of new homes in the United States have dropped further, and the stock of unsold homes has hit the highest level in ten years.[3] Not the bursting bubble that so many have been predicting, but definitely a slowdown.

Duckworth's sanguine forecast that sales at New Daleville would pick up in mid-January proves overly optimistic. In the six months since the sales trailers opened, NV has sold seven houses and Ryan only two. There has been no discernible pattern to the buyers so far. Ryan's sales representative, Kristi Oliveira, says that of the two homes sold, one is the least-expensive Sheldon, bought by a young family, and the other is a top-of-the-line model with all the extras, bought by an older couple. NV's first three buyers have been a single-mother nurse with a seven-year-old, a schoolteacher with grown children, and a middle-aged couple who are moving from an NV development in Virginia. "I've tried to make a profile of our buyers," says Karl Woodeschick, who is Oliveira's counterpart at NV. "There are empty nesters, young families moving up, and first-time buyers. I just gave up." He is referring to what economists call market segmentation—or, rather, its lack. One of the problems at New Daleville has been identifying potential buyers. Are they young families or so-called active adults? Are they willing to spend more for living in a neotraditional community—as TND advocates generally claim—or are they value-conscious and expecting to pay less? Without a definite profile, it's hard for the builders to target their marketing.

Oliveira remains upbeat. "The traffic at the site picked up in February," she says. "People like the concept, and they really like the houses. They think they're cute." Woodeschick is less sanguine. "There's a lot for buyers to think about," he says. "They have to buy into the location, the small lots, and the community association. It's a big commitment, and it usually takes them several visits to make up their minds. The problem is that it's too easy to just go home and say forget it." Both sales reps agree that there is very little overlap between visitors. Most people seem confused about whether there is one builder or two. "That's the first question they ask us," Woodeschick says. "Other than 'What are you guys doing way out here?'"

In real estate terms, Londonderry is the frontier. Prospective home buyers usually comparison-shop and visit several developments at once, but most of the new subdivisions in this part of Chester County are several miles from New Daleville. Only a fifteen-minute drive, but it could be another planet. It looks like Hovnanian's Brad Haber was right; it's not easy to sell houses on small lots in the middle of nowhere.

The April monthly meeting of the builders and the developer takes place in the Ryan sales trailer. Jason Duckworth and Jim Weidner attend, as well as staff from NV and Ryan. Although the chief activity on the site now is house construction, there are many outstanding improvements that are the responsibility of the developer. Weidner goes down a long checklist. He says that tree planting along the boulevard is progressing, and that bids are coming in for the "public amenities," referring to stone piers at the entrances, two arbors, a gazebo, and a windmill at the head of the boulevard, to create a rustic effect. Jason hasn't found an old mill and has had to settle for building a new one. There is a long discussion about the children's play lot. Arcadia suggests moving it to a more prominent location, in the wide median of the boulevard, but the builders object. Many of the buyers don't have children, they say, and won't see it as a plus. It is agreed that the play lot will be built in a more discreet location.

Weidner says streetlights have been ordered and will be up soon. The builders ask if they could be installed later, since on the narrow streets the extra-wide loads of the lumber trucks may knock them over. Weidner explains that the posts carry the street signs, which the township requires to be in place when the houses are occupied. "Your guys will just have to be careful," he says. Another signage issue concerns a warning to homeowners about the drip irrigation field. It appears that the lots abutting the field will each require a small sign. Jason is worried that their number will be excessive. "They'll be nice signs, and they'll say something nice," says Weidner.

Residential subdivisions are sometimes described as "springing up overnight." Hardly. It has taken New Daleville almost four years to get this far, and now, despite Arcadia's commitment and NVR's efficient organization, the messy business of preparing the site and building houses requires close attention to scores of small details. The contractors' pickups and vans parked on the narrow streets are in the way of NVR's bulky lumber trucks, which dump their loads of panels and wood trusses on the sides of the streets. Heavy cement trucks, pouring foundations and basements, create more congestion. This is especially a problem on the boulevard, which is where visitors to the model homes park. It's decided that signs are needed to divert construction vehicles onto a side street. It is also agreed that the builders can use two of the unsold lots to create temporary parking for visitors.

The discussion goes on for more than an hour. It is remarkably even-tempered, considering that the September deadline came and went more than six months ago and there is still no running water on the site. "There's a lot of tolerance on both sides," says Jason. "The builders have been understanding about the delays. On our side, even if the sales contract were in force, we would never oblige a builder to take lots without buyers. We all know that there are going to be bumps in the road."

Weidner reports that the water line has finally reached Daleville and that Chester Water is currently chlorinating and testing the system. He assures the builders they will have running water soon. This is a key concern, since the first four home buyers are slated to move into their houses in two months. Weidner explains that because of the length of the trunk line, to maintain pressure, Chester Water has had to build an intermediate pumping station. Construction of this booster pump is expected to take six months; in the meantime, the water company will install a temporary pump. Unfortunately, the resulting water pressure will not be sufficient for the fire hydrants to be effective. Weidner tells the builders that, as a temporary fire protection measure, Arcadia is proposing to the township to park a water tanker on one of the unbuilt streets. "Hopefully out of sight," someone remarks.

A few weeks later it becomes obvious that the estimates were too optimistic; the water system will not be operational for another month. And another complication has arisen. Londonderry's fire chief doesn't like the water tanker. He is worried that completion of the booster pump might be delayed into the winter, which means that water in an exposed tank would freeze. Weidner proposes an alternative: to prevent the water from freezing, Arcadia will bury a ten-thousand-gallon emergency tank in the boulevard median strip.

When the township granted final approval of the development plan for New Daleville, among eleven specific conditions was the requirement that there be a working fire hydrant within six hundred feet of any occupied building. Although there are no hydrants in any of the other subdivisions in Londonderry, the houses at New Daleville are closer together and represent a greater hazard. Technically, Arcadia is in violation of this clause, and the township could postpone the occupancy permits until the fire hydrants are operational. Occupancy permits, issued by the township building inspector, declare the houses to be legally completed. Without them, houses cannot be occupied, mortgage companies will not release funds, and builders cannot reach final settlement with buyers. Weidner believes that the township will approve the buried tank, since the fire department routinely pumps water from so-called fire ponds. I ask him if he thinks the water tank is the final hurdle for New Daleville. "God, I hope so," he says.

All these delays affect Arcadia, too. The contract calls for each builder to buy, or "take down," a minimum of four lots per quarter. However, until there is running water in the houses, not only are the builders under no obligation to buy a minimum number of lots but the annual escalation clause also does not apply. In effect, the builders are getting what Duckworth calls "a free ride." As usual, he is philosophical. "Dave Della Porta and Jason underestimated the problems we would have with Chester Water," he says. "Something unexpected always goes wrong in a project, but it's almost never the water. After all, that's just putting pipes in the ground. The truth is that we didn't pay enough attention to something that is routine." He estimates that, in terms of lost revenue, extra construction costs, taxes, and the carrying charge on the bank loan, the water problem could cost Arcadia as much as a million dollars.

At the May meeting, Weidner updates the builders. He says the fire chief has approved the buried tank, but the supervisors may need to take a formal vote at their next meeting, June 13. Bob Harsch, the township engineer, retired several months ago, and his replacement wants to do things by the book. Electricity is usually the last utility to be installed in a subdivision, after all other excavation is complete, and because of the delay in paving the streets, that operation, too, has been held up. The utility company is currently digging trenches and burying cables next to the streets. "How soon are they going to be done?" asks Bob Eager, who is NV's land manager and oversees its relations with developers. A couple of weeks, says Weidner. "I sure hope so," replies Eager. "We can't put the driveways in until they're finished. That's cutting it really close."

Timing has become a major issue for everyone. Both NV and Ryan have already canceled settlement dates with the first home buyers and rescheduled them for late June, which is only four weeks away. If all the pieces don't fall into place, the township won't issue occupancy permits, and people will not be able to move into their new homes. What happens then, I ask Oliveira. "Not much. Contractually, we have up to two years to deliver the house. But in the meantime we have to bear the carrying costs, which are ninety to ninety-five percent of the house price." Do purchasers ever walk away from their deposits? "I have heard of buyers walking away when their houses are not available on time, although it's never happened to me," she says. "But there's a lot of threatening, and it can definitely get unpleasant."

26

Hard Sell

How builders sell houses—traffic, qualified buyers, and guest cards.

Ever since Seaside, it has been common practice in neotraditional developments to build the first houses in a close group, in order to give buyers a sense of the neighborhood atmosphere that is TND's chief selling point. That is not happening at New Daleville. When I visit the site in May 2006, there are eight houses in various stages of construction. The four model homes are near the entrance, on both sides of the 150-foot-wide boulevard, but since they are so far apart, they don't really form a cohesive group. Two Ryan houses are going up in one far corner, and two NV houses in another. All are on edge lots. "We and NVR decided to give their sales personnel more freedom to sell around the site in a softening market," says Jason Duckworth, adding that, in hindsight, this may have been a poor decision. The scattershot appearance makes it difficult for people to see exactly what they are getting in return for putting up with smaller lots, back lanes, and narrow streets.

The model homes are fully landscaped, with grass, shrubs, and front walks, except for the Michener, which is still under construction. As soon as it's finished, Kristi Oliveira has been told to sell the Melville. "I wanted to keep it so bad, since half of my buyers are interested in the smaller models," she says. "But it was a corporate decision. They don't want more than one model home per development. I'll try to make an arrangement with the buyer so that I can take people through it from time to time."

The Carroll, which was to have been a model home, has been sold to a private buyer and is locked up, but the Melville is open, so I take a look inside. It is what builders call a white model, that is, undecorated and unfurnished. It really is white: Ryan paints the interior walls of all the houses the same color—called Cool Platinum—a builder's version of Henry Ford's famous pronouncement that Model T buyers could have any color "as long as it's black." But optional extras are much in evidence: the living and dining rooms that flank the foyer both have hardwood floors, crown moldings, chair rails, and bay windows, and the entry to the living room is accented by round columns.

The empty rooms are not large, but they feel spacious, thanks to the high ceilings. All the Ryan models at New Daleville have nine-foot ceilings on the first floor, and all except the Sheldon also have nine-foot ceilings on the second floor. Although American houses traditionally had tall ceilings, the earliest production builders, such as the Levitts, limited ceilings to eight feet. They did this to save money, but in the postwar period there was also a feeling that high ceilings were old-fashioned and wasteful—Wright's Usonians typically had low ceilings. Today, as a result of the popular interest in historic preservation and older houses, people have come to like the feeling of space, and buyers demand taller ceilings: nine feet has become the new standard, and ten feet is increasingly common in more expensive homes. The tall ceilings make the three bedrooms in the Melville feel light and airy. The master bedroom, or "owner's bedroom," as Ryan calls it, is even taller, open to the rafters with a so-called cathedral ceiling. Bedrooms have not changed much in design over the years, except that they are larger, a master bathroom has become standard, and there is more storage; in the Melville, all but one of the bedrooms have walk-in closets.

The basement is another part of the house that hasn't changed. The Levitts dispensed with basements to save money, but builders have long since discovered that, for the marginal added cost of excavation and concrete, the basement offers inexpensive extra space.* The basement in the Melville model home has a large finished space that most people would use for a children's play-room or a media room, an unfinished room that could be turned into a bedroom or a study, and an unfinished area containing the furnace and water heater. In addition, there is a half bathroom.

The focal point of the Melville, as of all Ryan houses, is the combined kitchen and family room that stretches across the entire rear. The carpeted family room has a gas fireplace and lots of light coming in through six-foot-tall, double-hung windows with low sills. The kitchen, with cherrywood cabinets and stainless steel appliances, includes an island and a sit-down counter. Extending from the back of the house is a low wing containing a laundry room and the morning room. The latter, with windows on three sides, feels like an old-fashioned sunroom. A door leads to a cov-ered porch and the backyard. The Melville has an unusual option: a finished room over the garage, reached by an exterior stair. The room is surprisingly large, about fifteen feet square, with an adjoining bathroom and a closet. It would make a nice guest suite, teenager's bedroom, or home office.

The two-car garage, like the sides and back of the Melville, is covered in vinyl siding, although the front of the house is brick, carried around the corner, following Mike DiGeronimo's sugges-tion. He has told me that he is pleased with the review process so far. "It's gone more smoothly than I thought it would," he said. "I haven't had any meetings about the houses with the township, but I've had many with the builders. I've been impressed by their con-cern to do things right. Most production builders just roll out their

---

*Basements are a standard feature of production houses in the Northeast and Midwest, less common in California and the South.

plans. I think that Ryan and NV are more creative in terms of design." What does he think of the houses built so far? "They've turned out pretty well. Even the dentil molding, which I was worried about, looks better in the field than it did on the drawings."

Ryan has included more intricate details than is usual for a production builder. The porch roofs are metal, and the porches have eye-catching Doric columns. The fronts of the basements, as well as porch piers, are covered in brick (the sides are concrete, molded in a brick pattern); the shutters are not operable, but at least they are sized to the windows; and the tall windows and protruding bays are generously dimensioned. The Melville, with its brick front, delicate dentil molding, white keystones, and arched Georgian entrance, looks a little too fancy for its rural location, but it has none of the flimsiness that bothered Tim Cassidy. This may not be authentic Chester County architecture—Ryan's advertising describes it as "Colonial-style"—but it is a cut above the plain-vanilla houses that most developers, including Ryan, build in this area.

Despite Cassidy's earlier objections, there is a lot of plastic. The porch columns are some sort of plastic, and the porch floors and steps are made out of a wood-and-plastic composite. The railings are vinyl, the windows are vinyl-covered, the shutters and trim are vinyl, and so is the siding. Builders like vinyl siding because it's easy to install and inexpensive—cheaper than stucco or wood, and much cheaper than brick. Unlike wood, vinyl doesn't require painting; unlike aluminum, it doesn't chip or dent. Most architects sneer at vinyl siding as inauthentic, but middle-income home buyers like it because it doesn't rot, isn't susceptible to termites, and never needs repainting. Vinyl siding has a useful life of about forty years, but the material has a major functional shortcoming— the color fades in sunlight. That is why standard vinyl does not come in dark hues. Buyers at New Daleville have a choice of white, cream, two shades of gray, and Desert Sand, that is, beige.

Ryan and NV have made an effort to improve their designs, but they are production builders, and it shows. The houses are handsome, but since they all come from the same Maryland factory, they are handsome in the same way. The details used by the two builders are similar, many of the materials are identical, even the windows are made by the same manufacturer. With so few houses, the lack of variety is not disturbing, but it will become more noticeable as the development fills up. When I ask Jason Duckworth about the houses, he admits that, in hindsight, a lot of the effort he put into the design guidelines has not borne fruit. "We're working with several different national builders, and NVR is by far the best, but they're still a bureaucracy," he says. "Many decisions can only be made by the head office. Since they're using stock plans, they've been able to make only minor adjustments, so the idea of having a local style hasn't really worked out."

I haven't spoken to Tim Cassidy for several months, so I call to ask him why Londonderry hasn't been more active in reviewing the houses. "The bottom line is that the design guidelines have no teeth, and Arcadia can, more or less, build whatever they want," he responds. "I also believe that we were given a false sense of security about the design of the houses. There were supposed to be several local builders to provide variety, but in the end what we got was Henry Ford. When I looked at those NV drawings, there were sixty sheets for a single house model, with all sorts of features that had nothing to do with New Daleville. That's when I realized it was over. My last-ditch attempt was to e-mail Mike DiGeronimo and put a shot across LRK's bows. But really, at the end of the day, there was nothing more I could do.

"People have busy lives and only so much energy," he adds. "At a certain point they have other stuff to worry about." The "other stuff" that has consumed the planning commission for the past year has been a controversial proposal to build a subdivision on a large tract down the road from New Daleville. The landowner has challenged the township zoning that prohibits community sewage, and after many heated public meetings—and a contentious supervisor election—the matter has ended up in court.

Cassidy remains critical of New Daleville's architecture. I think he was hoping for something less conventional; instead what he sees is the kind of commercial housing that he doesn't like. He thinks the way the brick wraps around the corners and suddenly stops is clumsy. And there's all that vinyl. "I guess I would have to say it's probably what I expected, because I expected to be disappointed," he concludes. "I know that the houses are not as bad as what is built by most developers around here, but I just wish they were better still."

Later, I ask Tom Comitta the same question. "I haven't attended any meetings in Londonderry since I last saw you at the plan approval meeting two years ago," he says. "In the previous neo-traditional developments on which I've worked, I've been asked to stay and oversee the implementation of the ordinance, making sure that the developer is doing things right. I was a little surprised that the township didn't ask me to get involved. So whether there's been smooth sailing between the township and the builders, I don't know." He tells me that he hadn't even seen New Daleville until a week ago. He's disappointed that the houses are so spread out and don't create a feeling of community. But on the whole, he says, he is pleased with the way it's turning out. "I know how hard it is to get these TND ordinances through. When I drove in, I thought, great. They finally did it."

Cassidy and Comitta are concerned about the appearance of New Daleville, but the dynamic of the project has shifted away from design guidelines and principles of traditional neighborhood development to something quite different: marketing. Kristi Oliveira, Ryan's sales rep, is at the center of this activity. She is a lively blond woman in her thirties, who has been with the company for seven years, although she started working at New Daleville only six weeks ago. Her enthusiasm for the houses is not merely that of a saleswoman. She and her husband live in an NV neotraditional house, which they built on their own lot outside West Chester last year (NVR sells individual homes without land to its employees, though not to the general public). "The neotraditional models, like the ones we're selling at New Daleville, are very different from our run-of-the-mill products," she says. "The special details really stand out. They're an easy sell. The hardest sell here are the tight spaces between the houses. People come to this area and expect large lots. We don't advertise the sizes, but they always ask."

The Ryan sales trailer is quiet. "Most people come on weekends," she says. How do prospective buyers learn about the development, I ask. They see the weekend directionals, she says, referring to the roadside promotional signs that are posted at intersections all over the township. Buyers also use the Ryan website to find houses based on location, price, and model type. They see advertisements for New Daleville in the local newspapers and read articles in new home listings directories, which are distributed free in sidewalk boxes. In addition, they may receive a flyer in the mail. Ryan's mass mailings usually target six to seven thousand likely buyers at a time, according to income, age, and family structure. The first mailing focuses on renters in the immediate area, people who have visited other Ryan projects, and owners of Ryan homes who might be ready to trade up. With feedback from this mailing, as well as information on who is visiting the sales office, future mailings are targeted more narrowly. According to Oliveira, the marketing effort for New Daleville has been greater than usual, which reflects the newness of the neotraditional product. There is no standard marketing budget, she says, in answer to my question. "Some projects don't need any advertising, some need a lot."

What about the slow sales at New Daleville, I ask her. "The

main reason is that traffic has been so light," she replies. Builders distinguish between "traffic" and all the visitors to a project. People tour housing developments for different reasons: boredom, curiosity, to get decorating hints. They are, in effect, window shopping. To identify serious prospects, or so-called qualified buyers, the sales reps ask visitors to fill in a questionnaire, officially called the Consumer Preference Survey but commonly referred to as the guest card. It includes personal information as well as the price range of the desired house and answers to questions such as "What are the most important features you want in a new home?" The card provides Ryan with statistics about potential customers and their preferences, but equally important, it is a sales technique. The assumption is that if people are willing to take the trouble to answer a questionnaire, they are more likely to be serious buyers. The back of the card contains the sales rep's follow-up record. "After the first visit, I'll make a thank-you call," says Oliveira, "and two or three more follow-up calls. Typically, on a second visit we'll walk the home site or do numbers. It usually takes three or four more visits before people come to a decision."

Oliveira says that there have been two kinds of qualified buyers at New Daleville: those whose chief concern is the price and quality of the house, and those who appreciate the neotraditional concept and are attracted by having a smaller yard in a walkable neighborhood. Many of the latter are what she calls active adults, that is, people with grown-up children. That is why Ryan added the Fitzgerald, which has a first-floor master bedroom suite and appeals to older buyers. Oliveira tells me that, because active adults are not concerned about school districts, they tend to look at many different locations, and since they already own homes, they're in no hurry and can take up to a year to make up their minds. This indecision affects her "conversion rate," which is the ratio of qualified buyers who actually put a down payment on a house. According to Oliveira, Ryan's national average conversion rate is 18 percent, although some of these sales are subsequently "lost," because buyers are turned down by mortgage companies, they can't sell their old homes, or they change their minds. This can happen often; both Ryan and NV have recently lost sales at New Daleville. Oliveira says that her conversion rate here is 13 percent, which is not disastrously low, except that the number of qualified buyers has been so small. It has been difficult to attract visitors without a furnished model. "I watch people drive into our parking lot, see the trailers, and turn around and leave," she says. "We're getting only six or eight qualified buyers a week. We'd like to see fifteen or twenty." She confirms that February to June is usually the best period for selling. That doesn't leave much time; it's May. When should Ryan start worrying? "About now."

Oliveira tells me that Ryan has been having success selling houses in a small subdivision called Country Walk, which is just behind New Daleville. "One of our models there is the Savoy, which is basically a Sheldon with a different exterior. It costs thirty thousand dollars more, but it sits on a much bigger lot," she says. On the way home, I drive over to Country Walk. The property originally belonged to the irate landowner who complained at the township supervisors' meeting about the proposed zoning change but whose development plan was subsequently approved. The subdivision consists of nineteen lots, arranged around a loop road. As at New Daleville, there are sidewalks, and half the site has been kept as a nature preserve with walking trails.* The similarity ends there, however, for the lots are large, between one-half and a full acre. There are currently a dozen houses in various stages of construction. Their architectural style is what builders call "contemporary," with fieldstone accent walls, oversize arched windows, and many overlapping gables. With room to spread out, most of the houses are larger than those at New Daleville, set well back from

---

*The Londonderry planning commission has insisted that the walking trails of the two developments be connected, a small but telling example of planning coordination.

the road, with long driveways, big yards, and plenty of privacy. Several have three-car garages. Altogether, an unremarkable development, but that may be precisely its attraction. I remember what Joe Duckworth told my students: "People want what everyone else has." That's what they get at Country Walk. The houses are selling briskly; according to Oliveira, only a handful remain unsold.

27

Competition

Why model homes are provided with plastic bundt cakes and memory points.

The Ryan sales office displays a plan of New Daleville covered with colored dots showing which lots have been sold, which are being held for potential buyers, and which are available. By late May there are four more red "sold" dots—three new homes as well as the Carroll model—and three green "holds." Sales reps sometimes add fake dots to encourage buyers, but Kristi Oliveira assures me that the new sales are real. Does this mean things are turning around? "I'm not sure," she says cautiously. "We now have more product to show, which is good. On the other hand, traffic is still light. We're not seeing many qualified buyers."

Dave Della Porta is worried about the slow sales. He has already given the builders a $5,000 concession on the price of the first lots, in exchange for a higher price later. Anticipating that NVR may demand to renegotiate the sales contract further, he commissions Brad Clason, a real estate consultant, to analyze the housing market in southern Chester County. Clason compares New Daleville's prices, features, and house sizes with those of fourteen other new subdivisions currently on the market. In the competing projects, lots vary from a third to one acre, and houses are three thousand square feet or larger. Builders in southern

Chester County are selling large houses on large lots—no surprise there. Prices are higher than in New Daleville, the average being about $385,000, compared with Ryan's $329,000; however, the per-square-foot selling price is actually less. In other words, the competition is offering buyers more house for their money—on larger lots and in less remote locations.

"You're introducing a brand-new small-lot concept into a rural housing market accustomed to one-acre lots," Clason reports to Della Porta. "With their smaller lot sizes, the New Daleville products should be positioned more competitively relative to the traditional single-family market in southern Chester County." Clason doesn't think that buyers here are ready to pay more to live on small lots. He recommends that Ryan drop its base prices by $15,000 to $20,000, and NV by $30,000. His analysis also reveals that, with the sole exception of an active-adult retirement community, the fourteen competing subdivisions are all selling fewer than two houses a month. No wonder Ryan and NV have been struggling—the real estate market in southern Chester County is currently stalled. It is unclear why. The national economy is strong, and although interest rates are going up, they are still reasonable. It may be that the three-year-old Iraq war has sapped people's confidence, or that all the talk of bursting housing bubbles has made buyers skittish.

Clason believes there may be another reason for the slow sales at New Daleville: the neotraditional concept is not well-represented. He does not mince words. "The site does not look good, the entrance is a mess, and the lack of a mass of completed homes in one section of the community prevents prospects from getting any sense of the curb appeal and lifestyle that a TND is all about." Normally, Arcadia would have taken care of these details, but the yearlong saga of the water supply, and the flurry of everyday crises caused by the subsequent delays, has diverted everyone's attention.

At the end of May, Rick King, NVR's land manager, and Mike Linthicum of Ryan ask Della Porta to reduce the price of the lots.

"In some of our earlier meetings I suggested that we might lower the price now in return for a higher price at the back end," says Della Porta. "But when we got the results of Brad's market analysis, I realized that was not going to happen. The market has definitely softened. We're going to reduce the price of the next ten lots, five each to Ryan and NV, by a straight fifteen thousand dollars per lot." In return, the builders have agreed to Della Porta's demand that the ten discounted lots all be on the boulevard. "Hopefully, by the fall we'll have enough houses in one area to create the buyer experience that is so critical to the TND concept," he says. "Sales will pick up, and we'll go back to our original prices." The builders have also agreed to Arcadia's request to emphasize the neotraditional lifestyle in their marketing. "It's a give-and-take," Jason Duckworth tells me. "This makes sense in terms of our relationship with NVR, since we want to be able to attract them to our other projects. And we definitely want New Daleville to be a success."

Joe Duckworth, who didn't take part in the negotiations but approved the new arrangement, takes the long view. "We've been caught in an economic cycle. It happens. It's obvious to me now that the housing market peaked last July, which was just when the builders opened their sales office. They set their house prices aggressively, assuming that the increases of the past year would continue, but instead of going up, prices went down, which put them way out of line. In addition, their costs have been going up in terms of building materials and interest rates, so their net yield is lower. They're getting squeezed."

On the boulevard, the selling price of the Sheldon will be $20,000 less, which means that New Daleville now starts in "the mid-$270s," rather than "the upper $340s," the price nine months ago. The price of the Melville is reduced by $10,000, and the Fitzgerald by more than $40,000. Oliveira is pleased, not only about the reductions but also about the prospect of having a group of houses on the boulevard. "If we can just show people how it's going to be, it'll make such a big difference," she says. What about the buyers who have bought houses already, I ask her, remembering Duckworth's admonition about not lowering prices once a project is under way. Won't they be upset? "It's not something I'm going to tell them, but they may see our ads. We haven't sold any lots on the boulevard, so it's not a direct issue, although they could say that they would have bought one of these lots if they had known. We're definitely not going to give them a discount, but I don't want them to feel badly. It's going to be a difficult conversation." In fact, only one owner notices the price change and complains. He is satisfied after learning that the reductions will be limited to the lots on the boulevard, and that the prices of the unsold edge lots, like his own, are now slightly higher. For Ryan, increasing the premium on the popular edge lots not only mollifies the previous buyers but also slightly offsets the lower prices on the boulevard.

A few weeks later, Oliveira gets her furnished model. The Michener will be Ryan's largest house in New Daleville, more than 2,500 square feet. Fitted out with all the options, it sells for more than $400,000. A model home is a compromise between an idealized dream house and something realistic enough that the prospective buyers can imagine themselves actually living in it. In the Michener, the rooms are furnished and decorated with small touches to make them appear occupied: area rugs, floor lamps, framed pictures on the living room credenza. The illusion is reinforced by a dining room table set for dinner, complete with candles and a floral centerpiece. The feeling of a stage set continues in the butler's pantry, where a granite serving counter holds a coffee set and an iced bundt cake. In the kitchen, a cookbook is propped open to a recipe for onion-zucchini bread next to a mixing bowl filled with batter and a cutting board with—two onions. Like the bundt cake, the batter and the onions are convincing plastic imitations, as realistic as the sushi samples displayed in some Japanese restaurants.

The imaginary owners of the Michener seem to sit around a lot: the living room, the family room, and the morning room are all furnished with sofas and easy chairs. The furniture is traditional—no black leather and stainless steel here—without being overtly old-fashioned, the sort of thing one sees in catalogs from Crate & Barrel and Restoration Hardware. The bedrooms continue the comfortable theme. The walls of one are painted a cheerful apple green with stenciled flowers—it could belong to an older daughter; the second has a tray with a coffee cup and a pecan tart on the bed and appears to be a guest room; the third is Oliveira's sales office. The hall, with a varnished wooden floor, chair rails, and crown moldings, overlooks the foyer below. Since New Daleville is still waiting for its water connection, the fixtures in the hall bathroom, as in all the bathrooms in the house, are inoperable; a colored ribbon is discreetly tied across the toilet seat.

The master bedroom is entered through double doors. A king-size bed is piled high with pillows, shams, and throws in various shades of brown to complement the hand-painted striped walls. (What do people do with decorative pillows at night? Pile them on the floor?) The room is huge—the largest in the house—with space for a sofa and coffee table at the foot of the bed, facing a flat-screen television mounted on the wall. The owner hasn't been watching a daytime soap, though; a pair of reading glasses and a half-open book lie on the bedspread.

The book is Susanna Moore's *In the Cut,* an erotic thriller. The unlikely title is an anomaly—the Pittsburgh firm that decorates NVR's models buys used books in bulk. Which is not to say that the Michener's décor isn't designed with a particular customer in mind. "We thought that the buyer would be somewhere in between a traditional and a contemporary," according to Ryan's Carmela Bond, citing common marketing categories. "Trendy at a moderate price." That explains the comfortable, middle-of-the-road furniture and the colorful but unostentatious accessories.

The trendy part of the Michener is in the basement, which is furnished as a game room, with a trio of framed Indiana Jones movie posters and a large sectional couch facing a television cabinet. Beside the sitting area is a baize-covered pool table with racked-up balls. Like the long, built-in, granite-topped bar, it is what model-home decorators call a "memory point," a dramatic feature that is intended to catch visitors' attention and differentiate this from other model homes they may see while house hunting. Beyond the bar, at the far end of the room, is the so-called sample area, which includes a mock-up of a kitchen cabinet showing various wood finishes, a display of different brick, siding, and shutter samples, and an entire wall devoted to bathroom tiles and swatches of carpeting and vinyl flooring. After the carefully arranged décor, it's like going backstage.

28

The Spreadsheet

Even though production models are standa
exterior materials, siding and trim colors, in
and various options, the buyer of a typical n
make fifty to a hundred individual choices.

Scott and Meghan Andress are among the firs
Daleville. They currently live in Sadsburyville, a
away, in a three-year-old Ryan town-home developm
I visit them in April. The hundred houses in their developi
arranged in groups of four, five, and six. The façades are
stucco, and stone, but the variety only emphasizes the fact that i.
houses are basically the same. There is not much landscaping. It is
a classic starter community, where architectural niceties are sacri-
ficed for the sake of low prices, enabling first-time buyers to get
a toehold in the market.

I talk to Meghan in her kitchen—Scott is at work. The rooms
are mostly empty, the furniture in storage, since the house has been
sold and settlement is in about two weeks. She says that they will
move to an apartment. "We don't yet know exactly when our new
house will be ready. We hope June or July. It depends when they
get the water connected. I've also heard that there is some problem to do with an emergency water tank."

A tall, freckled redhead in her late twenties, Meghan grew up

...ginia, and studied forestry at Virginia Tech,
...tt, who was an engineering student. He was
in Willia... ...Chester County, not far from Londonderry, and
where ...a local consulting firm. They've been married
bo... ...ave a fifteen-month-old daughter, Kendal.

...a project manager for an environmental consulting
...Phoenixville, a forty-minute commute. She is on
...ot—her company pickup is parked in the driveway—
...versees the construction of sewage treatment systems
...southeastern Pennsylvania and southern New Jersey.
...ly, the fact that there was a community system rather
...dividual septic tanks was one of the things that attracted us
...w Daleville," she says. "Since I design drip irrigation for a
...g, I understand how it works."

...he Andresses' town home is twenty feet wide, with a garage
...d playroom on the first floor, living room and eat-in kitchen on
...he second, and three small bedrooms on the third. There is an
eight-foot-square deck off the kitchen, above a similar size patio.
"We bought the house brand-new, three years ago, thinking we
would keep it for two to five years," she says. "Now we want a
proper yard and a bit more space. We're ready to move." They sold
their house for $215,000, an increase of $65,000 over what they paid.

What attracted them to New Daleville, I ask. "We like Londonderry because it's in the Octorara school district. The district
where we live now is not great, and if we didn't move we'd have
to send Kendal to a private school." Was the relative remoteness
of Londonderry an issue? "It means a slightly longer commute for
both of us, maybe ten minutes or so. There's a good supermarket
seven minutes from New Daleville. Exton Mall, where we usually
do our larger shopping, is about half an hour away, which is a bit
far, but since we both work, we tend to stop there on the way
home, so that's not really a problem."

I ask her how they heard about New Daleville. "Last October,
we got a flyer in the mail, offering a thousand-dollar credit to Ryan

homeowners who bought there. We were thinking of moving, and we were pretty happy with our house, so we liked the idea of living in another Ryan development. The problem was the price. The cheapest houses were three hundred and twenty thousand, which seemed high to us since this was far out in Londonderry. So we didn't bother following it up."

They forgot about New Daleville until three months later, when Scott's mother called and said that she had been researching new developments in Londonderry for his sister, who was also looking for a new house. She had investigated Country Walk and New Daleville. "Your sister has decided to buy in Country Walk," she told him. "She likes the large lots and the privacy. But I can see you and Meghan in New Daleville. Did you know that they've just reduced their prices?"

"That's when we decided to take a look," says Meghan. "One of the things we like about where we live now is the closeness of the community and getting to know our neighbors. So the village arrangement appealed to us." Had they heard of traditional neighborhood development, I ask. "I wish I could say we had, but we were unaware of neotraditional planning and new urbanism. Of course, Oxford, where Scott grew up, is a small town, so that was familiar." I mention that Colonial Williamsburg, near where she grew up, is one of the models for neotraditional planning. "The part of Williamsburg where I lived wasn't like that," she replies. "It was very spread out, with two- and three-acre lots and gated communities."

How often did they visit New Daleville before making up their minds? "Once a day for two weeks," she says. The Andresses give new meaning to the term *conscientious consumers*. They looked at different lots. They took a measuring tape and string and staked out the house and the perimeter of the lot. They focused on the lots at the edge of the site. "We didn't like the back lanes," Meghan says. "We have three vehicles, my big company pickup, my Jeep, and Scott's Buick, so we knew that parking was going to be a problem. We chose the option with a detached garage behind the house, which gives us a nice long driveway."

After all Joe Duckworth's strategizing, Bob Heuser's ingenious planning, and Tom Comitta's and Jason's efforts to follow neotraditional orthodoxy, it has come down to this: where do we park? Despite the sensible arguments in favor of small lots, narrow streets, walkability, and density, buying a house is not, for most people, about ideology. It's about fulfilling personal dreams and practical needs: Do I really want a Jacuzzi in the master bathroom? Are the kitchen cabinets large enough? Would I like a front porch? And, at least for Meghan and Scott Andress, is there sufficient parking? Negotiating large vehicles through narrow rear lanes didn't appeal to them. Most Americans park in their driveways, either because they use the garage for storage or because the driveway is more convenient. There is no space to leave your car outside in the back lanes at New Daleville, so those owners will have to park inside their garages—every time.

The Andresses settled on a lot facing the small cul-de-sac that Bob Heuser had added at the last minute, to accommodate the redesigned drip field. "We did consider a lot facing the street," Meghan says, "but we didn't like having traffic in front of the house, and worrying about Kendal wandering into the street." She points out that their house in Sadsburyville is also on a cul-de-sac.

What about the size of the lots at New Daleville? "We weren't too worried about that," she says. "Our lot here is only two thousand square feet, so they were a big improvement. Since we both work, we didn't want to have to take care of a large backyard." However, they were concerned about the closeness of neighboring houses. "We didn't like the regular lots, where the houses are about twenty feet apart." That was the other reason they chose a cul-de-sac lot. Because of the side driveway and the pie shape of the lot, there will be about eighty feet between them and their neighbors. "We spent a lot of time making sure exactly where the houses next to us would be located."

The lot that the Andresses chose is actually the largest at New Daleville, almost a quarter acre. "Ryan was charging a sixty-five-hundred-dollar premium for these edge lots," Meghan says. "But when we went over the plans, we discovered that there was a nine-foot drainage easement on one side, where we couldn't plant or put in fencing. So we were able to get them to lower the premium to fifteen hundred. We were the second buyer, so at that point they were willing to do just about anything."

The Andresses had bought their Sadsburyville town home before it was built, so they had experience working with a builder, choosing options and upgrades. Even though production-builder models are standardized, between color choices (at New Daleville there are six vinyl siding colors, four shutter colors, and three trim colors, depending on the model), the interior upgrades, and the various options, the buyer of a typical Ryan house has to make fifty to a hundred individual choices. Scott and Meghan went about it methodically. They looked at the plans and fired off almost daily e-mails with long lists of questions for the Ryan sales rep—Oliveira's predecessor. Is there one heating zone or are there two? (One.) Can we add a laundry chute? (No.) What is the material of the front walk? (Brick.) How many overhead bulbs are in the basement? (Not sure, we'll check.) "We looked at the Sheldon, the Melville, and the Austin, comparing the base prices and the cost of combining various options," Meghan tells me. "We made a *lot* of Excel spreadsheets."

They chose the least-expensive Sheldon. "When we built our Sadsburyville house, we knew we wouldn't be in it very long, so we didn't spend money on a lot of extras," says Meghan. "But we plan to be in our New Daleville house for a while, and we wanted a model that we could afford to upgrade, with better cabinets, hardwood floors, recessed lighting, things like that." Their town home has a sunroom off the kitchen that they like, so they chose the morning room option, as well as a two-foot extension to the family room. They added doors to the living room, which they plan to use as a study. They dispensed with fireplaces—"we have neighbors that have them and just don't use them"—and bay windows. They decided to leave the basement unfinished. "We want to install lots of built-in storage, so we'd rather finish it ourselves." Although Mike Linthicum had told me that each model had a limited range of specific options, in exceptional cases, Ryan will accommodate what they call a nonstandard option, assuming the change can be easily made and the buyer is willing to pay. The Andresses wanted to substitute a shower stall for the tub in the master bathroom, which cost them $1,200. Altogether, the various options added $45,000 to the base price of the house and increased the total area to more than two thousand square feet (their present home is sixteen hundred square feet). After they signed the contract, Ryan introduced its mortgage incentive plan, so they took that, too, insisting on the $15,000 bonus options. Which extras did they add? "We got better countertops, a vaulted ceiling in the master bedroom, upgraded cabinets in the bathrooms, and a carpet upgrade," Meghan says.

In her first e-mail of questions (thirty-five of them), Meghan asked for a copy of the homeowner association agreement. "We read it cover to cover," she says. "Our town home development has a homeowner association, and we got caught off guard because we hadn't read the documents closely. It turned out that you needed permission to extend the flower beds in front of your house. We applied to the architectural committee. What was frustrating was that we got turned down, while other people just went ahead and did it, ignoring the rules." Meghan is critical, but she sees community associations as a necessity. "I wouldn't want to serve on the board, but we go to the meetings and participate." The fee at New Daleville will be $130 a month, more than double what the Andresses pay now. "I want the public gardens and park space to be taken care of, and that costs money," says Meghan. She is happy that the children's play lot will be built soon at New Daleville. "They promised us a play area here," she says of their present development, "and in three years they still haven't built it."

As a home buyer, Meghan deals exclusively with the builder. She is aware that there is a "town architect," who has reviewed the design of their house, but from our conversations it is obvious that she associates him with the township, not with the developer. She likewise credits Londonderry with requiring tree planting and landscaping in the public areas. Although the subdivision is largely the result of Joe Duckworth's vision—in that sense he is the founder of New Daleville—at this point in the process, the developer has become largely invisible.

By early June, the Andresses' house is nearing completion. The Sheldon, like the Melville, is a simple box. Covered in light gray siding, with a deep front porch, it resembles a farmhouse. The porch roof is propped up by two-by-fours, awaiting the delivery and installation of permanent columns. It's raining lightly as I pick my way across the muddy front yard. This is still a building site. A Dumpster full of construction debris and a portable toilet are parked in front.

Scott Andress is waiting for me on the porch; he's a small man in his early thirties, with a brush cut. He's wearing construction boots. He and Meghan have been visiting the house every few weeks. We're joined by Greg Norbeck, Ryan's project manager, who is responsible for supervising construction. The first thing that Scott asks him is if there is any news about the water connection. Without water there can be no settlement, and the uncertainty about exactly when the house can be occupied has caused Scott and Meghan to delay signing their mortgage. Signing locks in the interest rate for sixty days—if the settlement date is postponed beyond that, the guarantee expires. They are nervous, since rates have been creeping up, and yesterday, unable to wait any longer, they signed. "Don't worry," says Norbeck. "I'm sure we'll have the water connected in a week or two. Then we'll install the carpeting. We've tested the plumbing system with compressed air, but I don't like to put in the carpet until we have running water, just in case. If there's a leak, it's bad enough replacing Sheetrock." He adds that the township has approved a buried emergency water tank for fire protection, but he seems unsure about exactly when it will be installed. In any case, he's scheduled the settlement for June 23, three weeks from now.

Meghan is the last to come in out of the rain. "Greg, tell me that the front porch is leaking because the roof isn't finished," she says. He confirms that they're waiting for a special order of aluminum roofing, to match the dark green color the Andresses chose for the shutters. "I hope it's as good as the regular material," she says. "Actually, I think it's more expensive," he says. "Great, great," she replies. She has also noticed that the porch ceiling is buckling in places. Norbeck says it must have been improperly installed. "I'll make sure it's redone," he promises.

We go upstairs. Meghan points out that the ceilings are a slightly different color than the off-white walls. "When we bought the house, that was standard. Now they paint the walls and ceilings the same color to save money, and it costs extra to have white ceilings." She says that they had Ryan reverse the swing of the master bedroom door so it wouldn't interfere with the bathroom door, but now she notices that the light switch is in the wrong place. It will have to be moved. The master bathroom is small but airy, thanks to the extremely tall ceiling. "I like it, too," she says. "When we got the cathedral ceiling in the bedroom, we didn't realize that it included the bath." Scott has noticed that there are irregular cracks between the window frames and the wall opening. "That'll get caulked," Norbeck says, "you won't notice it." There's a lot of caulking in the house, since the standard of on-site workmanship is not high. I can see why Ryan tries to get as much done in the factory as possible.

Meghan says that she would like to vacuum the floors before the carpet is installed. "Don't worry," says Norbeck, "I'll make sure it's clean." As we go down the stairs, Meghan asks if the single light fixture will be sufficient to light the staircase. "I think it should be enough," Norbeck says. "But we can always add another one. It's not a big deal."

Meghan spots an oddly located towel bar in the powder room, and sure enough, when Norbeck brings a set of construction drawings from his truck, the bar turns out to have been mounted on the wrong wall. "We'll get that moved," he says. The inspection is a series of small negotiations. Sometimes the builder has missed something, sometimes the buyers change their minds. Sometimes one side prevails, sometimes the other. Sometimes it's nobody's fault.

The kitchen is nice though hardly fancy, with dark wood cabinets, Corian counters, and a vinyl floor. The view out the morning room windows, however, is spectacular: rolling fields, clumps of trees, Charlotte Wrigley's barns and silo in the distance. When I mention it to Meghan, she says, "Isn't it great? As soon as we saw it, we put down money to reserve this lot."

Scott has noticed that one edge of the sit-down countertop is not beveled to match the rest. At first, Meghan is not convinced that it is worth complaining about. "We have other battles to pick," she tells him. But as she looks longer, she changes her mind. "Do you see it now?" he asks. "Yes, I do. Definitely." Norbeck adds another note to his thick binder.

There is a three-foot drop from the outside door in the morning room to the ground. Meghan asks about steps. Norbeck isn't sure, since the construction drawings don't show any (many owners add a deck or a raised patio, so Ryan typically does not provide steps). She points out that there is a concrete path leading to the door, and Ryan has installed a lockset on the door at their request, so this door is definitely an entrance and must have steps. Norbeck agrees and says that they'll be either concrete or Trex, referring to the synthetic wood material that is used for porch floors.

"Oh, definitely Trex," she says. "Please." He hesitates. "I'll just tell them it has to be Trex," he says.

Norbeck says the rain has delayed putting in the driveway, but he expects that to be done early next week. Meghan asks about the landscaping. "We put in sod over the whole lot," he says. "We usually try to do that as close to settlement as possible." Meghan is concerned about watering. "I could bring some hoses and stuff over here," she says. "I'd hate for it to die." Norbeck says that his people will do it but quickly adds, "There's no guarantee on the sod. If a patch dies, we'll replace it. But not if your entire lawn dries up and your neighbor's is green."

Building a house, even a highly standardized and rationalized "product" such as a Ryan home, is complicated. Construction starts with excavation and poured concrete and ends with carpeting and light switches—and sod. These bits and pieces, crude as well as delicate, must be brought together, if not seamlessly, at least smoothly and quickly. That is why housing has confounded generations of advocates of industrialization and mass production. At one point, all house construction is individual: this house, this place, these people. The Sheldon model was designed by NVR's architectural division two years ago, but *this* Sheldon is the result of what is happening here, today. It needs someone like Greg Norbeck to coordinate the work of the many different trades. It needs vigilant buyers. It needs nitpicking, adjustment, and smoothing out to get it right.

"Cookie-cutter," snobbish critics call builder houses, but the Andresses' future home reflects personal involvement in numerous small ways. It's not the same as having a house designed expressly for you, of course. But there are no cost overruns, no nasty surprises, and I must admit, no headstrong architects to confront. I've designed houses for individual clients, whose participation was necessarily great since we were starting with a blank slate. Still, it is obvious that Meghan and Scott are caught up in the design and construction of their home. When they move in—they hope in a few weeks—this will definitely be *their* place.

We leave the house and stand on the half-finished porch. It's stopped raining, and the sun is out. Meghan turns to Scott. "Are we happy?" she asks. He nods. Yes, yes, they are.

Two and a half weeks later, I meet Jim Weidner, Arcadia's construction manager, on the front porch of the Michener model home and ask him about the water situation. He tells me that testing and adjusting water quality has taken longer than expected, but that he's been promised the finished houses will have running water in a day or two. New Daleville must also have fire protection, but I haven't seen any sign of a water tank. What's going on, I ask Weidner. He rolls his eyes. "The supervisors approved a buried tank. We were supposed to borrow a steel tank from a company that we do business with, but at the last minute the owner got worried about liability and said we had to buy it. He was asking too much, so I found a concrete tank that we could leave in place once we didn't need it anymore. The township engineer has been pretty adamant about the need for a buried tank, but I thought it wouldn't hurt to ask one last time. Since the construction of the booster pump was going so well, I said, and it looked like it would definitely be finished by August, did we really need to bury the tank? 'I guess not,' he said. I guess not! So now we're renting a standard steel water container. It looks like a small Dumpster. I expect delivery at the end of this week, on Friday morning, which will allow the builders to settle their first houses in the afternoon." Just in time.

From my vantage point on the porch, I look over the rest of the site. The neat picket fences, trimmed lawns, brick walks, and carefully tended shrubbery of the model homes only serve to accentuate the generally untidy appearance of the surroundings.

The grass on the unbuilt lots is getting longer. Red earth is piled up around freshly dug utility trenches, and orange and black telephone and cable conduits stick out of the ground at intervals. Building materials are piled untidily along the road beside two unfinished basements, which now makes eleven houses in various stages of completion. When I go over to get a closer look at the Andress house, there is a painter's van parked outside. The porch roofing has been installed, and carpenters are putting up the columns. But the driveway is not done, and neither are the sidewalk, the front path, or the landscaping. It's pretty obvious that the house won't be ready for June 23, which is only three days away. When I ask Meghan about this later, she says that, despite the water problems being solved, for some reason the work on their house has been delayed, and their settlement date has been pushed forward to July 13. They plan to move the following day.

29

Moving Day

After all the houses have been sold, New Daleville will represent a total investment by buyers and lenders of about $40 million. All this is the result of an initial private investment of $2.5 million.

When George Washington was seventeen, he worked as an assistant surveyor on the new town of Alexandria, Virginia, a no-nonsense subdivision of eighty-four half-acre lots. The only fancy touches were the street names: King, Queen, Prince, Princess, Duke, Duchess.[1] Street names are a chance for the developer to set a tone for a project—and to leave his mark. In his plain fashion, William Penn named the east-west streets of Philadelphia after trees. Henry Howard Houston renamed several of Chestnut Hill's numbered streets after Indian tribes: Seminole, Navajo, Cherokee. For New Daleville, Jason Duckworth has chosen the names of early Londonderry settlers—McGrew, Robinson, Neill. The main thoroughfare will be Wrigley Boulevard, to commemorate the owners of the family farm that once occupied the site.

Scott and Meghan's house is on Columba Street, named after the patron saint of the original Londonderry, in Ireland. I visit them on July 16, two days after their move. It's an imposition to show up so soon, but Meghan has assured me that it's all right.

Throughout this long project I've been looking forward to the day when a family actually moves into a house. Now that it's happened, it seems like an anticlimax. There should be a ceremony, like a ribbon cutting or a ship launching. All the people in the New Daleville story should be here: Tom Comitta, to give a homily on neotraditional neighborhoods, Joe and Jason Duckworth, Dave Della Porta with his pro formas, Bob Heuser carrying a roll of plans, Mike DiGeronimo and his design do's-and-don'ts, Tim Cassidy and the planning commission, the Londonderry board of supervisors, Mike Linthicum leading the Ryan team, as well as a group of out-of-town visitors, the Seaside contingent, Robert Davis, Andrés Duany, and Melanie Taylor. All of them lining Wrigley Boulevard, applauding Scott and Meghan as they move into their new home.

Instead, there is just the same little house, whose construction I've been following, off and on, for the last three months. Now it's finished: porch, front walk, driveway, shrubbery, and lawn (rather burnt-looking since we've just come through a heat wave). Finished but different. There are cars in the driveway, Meghan's Jeep with a small trailer attached, two aluminum lawn chairs and a small child's chair on the porch, a rug drying on the railing, a doormat and a scuffed pair of sneakers. The previously empty windows have Venetian blinds. Through the open garage door I can see piles of cardboard boxes. In other words, there are signs of life. It's as if someone had flipped a switch, from "House" to "Home."

The interior doesn't look anything like the neatly furnished Ryan model, but I expect it never will. It's not just the disorder of the move. The model didn't have a colorful plastic slide in the middle of the family room, over which Kendal, a happy little girl with curly red hair, is clambering. Her toys are everywhere. "They've been in storage for the last two months, so when we unpacked it was like Christmas," says Meghan.

She is putting away dishes and utensils in the kitchen. Scott is helping his father hook up the dishwasher. "We thought that the

Ryan appliances were pretty expensive, so we got everything from a Scratch and Dent in Delaware," he says. "The fridge is coming on Tuesday, so for now all our food is stored at the neighbors', who moved in just before we did." Meghan and Scott have already met the people in the house next door, as well as a young couple whose house is under construction on the other side. "You can't ignore the neighbors," says Meghan. "After all, you have to be a sociable person to want to live here. We noticed that the people down the street have little kids. We'll probably walk over there this week and introduce ourselves. Friends for Kendal, friends for us." For the moment, only four houses in New Daleville are occupied, but already a little community is beginning to form.

Meghan's parents, who have just left, came up from Virginia to help with the move. Her father, as practical as his daughter, installed the blinds as well as ceiling fans. "Our first night here, the day we moved, was cool and breezy," she says. "We all sat out on the porch. It was wonderful. There are no streetlights yet, and we could see clear sky and bright stars. And it was so quiet. The others finally turned in, but my father and I stayed out there a long time."

Meghan had stored some of her furniture, as well as her piano, with her parents, who brought it up in a U-Haul, along with other things. "They're downsizing at the same time as we're upsizing," she says. "So we're getting all their stuff." Exactly, I think, as I leave the house. As my chapter of the New Daleville story is drawing to a close, theirs is just beginning.

New Daleville is a classic tale of entrepreneurship, of risks and opportunities, plans and mishaps, ups and downs. Private money got the ball rolling. Despite unforeseen stumbling blocks and delays, the developers managed to transform a cornfield into permitted lots worth about $15 million. The home builders, at considerably less risk, are marketing the houses. Buyers have come—or not come—by individual choice. National economic forces have intervened. Prices have been modified, incentives offered, deals struck. The builders will make money, too, though not as much as in the heady years of the housing boom. They will also learn important lessons about marketing a neotraditional development: build the model homes early, create a sense of community as soon as possible. Two or three years from now, when all the houses have been sold, the development will represent a total investment by buyers and lenders of about $40 million. Based on historical experience elsewhere, this investment will increase in the future, to the benefit of the individual homeowners. All this is the result of an initial private investment of $2.5 million.

Of course, New Daleville is not simply about money, it's about creating a community. When it's complete, this neighborhood will not have the variegated appeal of a mature garden suburb such as Chestnut Hill. The trees will have to grow, the landscape will have to fill in, and the houses will have to develop the scuffed edges and small adjustments that give a place character. Vinyl doesn't age as well as Wissahickon stone, but when it is eventually replaced, there will inevitably be more variety. At some point a house will be enlarged; another will be modified. Just as the Levittown houses sprouted shutters and dormers, the New Daleville houses will sprout—whatever is in fashion in forty years. The homeowner association will try to maintain uniform standards, but individuality will creep in, as it always does.

That is the lesson of Chestnut Hill and Levittown: community building takes time. Which is not to say that it doesn't greatly matter what the developer builds in the first place. The architectural quality of George Woodward's houses in Chestnut Hill has had long-lived appeal. The flexibility of Alfred Levitt's designs has allowed future generations of homeowners to adapt and expand. At New Daleville, the public spaces will always serve to tie the houses together. The architecture may change, but the feeling of a neigh-borhood will endure. In all cases, the initial plans were only the first step in a long process: from developers to builders to home-owners.

Americans take this process for granted, but there are other ways to build communities. Some countries depend on centralized planning, in which the public authorities decide where people should live, and what kind of housing they should live in. Others rely on private builders but are strict about controlling the avail-ability and use of land. The cities of the developing world, by contrast, depend on the unplanned and unregulated efforts of mil-lions of individuals who build their own homes in so-called squat-ter or informal settlements, which eventually turn into urban neighborhoods.

For better or worse, America has always approached commu-nity building as a business. Developers and builders have proved proficient at providing the kind of housing that people want, at prices most of them can afford. Over the years, what people have wanted has changed, as reflected by the early garden suburbs, the mass-produced postwar subdivisions, and the master-planned communities of the prosperous decades that followed. Picture windows and carports one decade, dormers and porches the next. Today's neotraditional developments are more than nostalgia, however; they are an expression of a taste for design, neighborli-ness, and sociability. What does the future hold? It could be a con-tinued dispersal to the edge in yet farther-flung exurbs, some version of Wright's Broadacre City, or a densification of the older suburbs to make them more townlike, more Unwinesque. Or likely both.

The market approach to community building has proved to be myopic as well as accommodating. Each development project is an isolated business venture—a New Daleville here, a Country Walk there—with little coordination among them. As Londonderry shows, sometimes the parts come together to make a whole, sometimes they don't. Finding ways to establish and reinforce links among different developments remains a major challenge, for municipalities as well as for developers. In this process, places where land use is controlled by county governments will be better equipped to develop coordinated plans; in those parts of the country governed by fragmented home rule, piecemeal growth will likely continue.

The story of New Daleville is a small episode in the ongoing development of the United States, and it is repeated, in different versions, across the country. Joe and Jason Duckworth take their places beside the generations of land jobbers, speculators, subdividers, and real estate developers who have built communities across the nation, in cities and towns, suburbs and exurbs. Directly and indirectly, these entrepreneurs have shaped the way Americans live today—and will live tomorrow.

POSTSCRIPT

New Daleville, September 2006

The housing market slowdown that began in the summer of 2006 continues into the fall. A variety of factors—uncertainty about interest rates, buyers' concerns about the future, and house prices that have simply risen too high—has caused the market to stall. Home sales are dropping nationwide, and people who can't sell their old homes are not in a hurry to buy new ones. Builders' inventories grow. The prices of new houses stop rising in some regions, and are declining in others. While this is hardly a bursting bubble, the peak of the real estate boom is clearly over.

The first to bear the brunt of the downturn are the national builders, whose stock prices plummet. Corporate belt-tightening quickly follows. At New Daleville, having already cut prices, the builders take a second look at the models themselves. Certain special features that were included when the houses were first marketed—brick pathways, varnished front doors, nine-foot ceilings in the bedrooms—become extras. Even the trees that come with the houses are now slightly smaller.

Although projects such as Celebration and I'On have shown that buyers will sometimes pay more to live in a well-designed community, neotraditional developments are not immune from economic downturns. Nevertheless, despite the fact that August is traditionally a slow month, with families on vacation rather than house hunting, traffic at New Daleville actually increases, and continues into September. In two months NV and Ryan each sell four more houses, for a grand total of twenty. A local builder buys Arcadia's five remaining house lots, as well as the commercial lot. It appears that the new strategy is working—at more competitive prices, New Daleville is finding buyers. It also helps that the marketing now emphasizes the community aspects of the development. Jason Duckworth and his assistant, Christy Flynn, are spending several hours each Sunday at the site, explaining the neighborhood concept to the sales reps—and to the buyers. Flynn brings her laptop and shows images of other neotraditional communities. Now prospective buyers hear not only about numbers of bedrooms and kitchen options, but also about public parks and walkable streets.

Tom Comitta suggested to Jason that the sales reps might benefit from more exposure to neotraditional ideas. Comitta also had thoughts about improving the boulevard by planting more London plane trees and adding a small gazebo (which Arcadia does). "The public realm must be implemented with the same attention to detail as the houses," he tells me. "What's needed is a town *landscape* architect." Comitta is no longer working for Londonderry Township, but he remains interested in New Daleville for professional reasons. His practice has grown; in addition to advising municipalities, his office is now designing several subdivisions. Needless to say, they are all neotraditional. New Daleville hasn't changed any of Comitta's ideas about traditional neighborhood development. In his thoughtful way he is persistent. While acknowledging that planning theory sometimes comes up short against the realities of the marketplace, he remains as enthusiastic as ever about compact, walkable neighborhoods.

So far, with only ten houses occupied, New Daleville doesn't look much like a neighborhood. Yet the invisible bonds of community are already beginning to form. According to Meghan Andress, her neighbors are friendly and outgoing. Since people are moving in one at a time, she meets them all. "We usually go for walks in the evening. People are out on their porches and we start talking," she says. "They invite us in for a glass of wine. It's neat." Meghan's daughter, Kendal, is going to swimming lessons with a new friend; Meghan and a neighbor down the street are planning a community cookout for the fall. How much this activity is encouraged by the proximity of the houses and the presence of front porches is diffi-cult to say, but it's hard not to believe that people are affected by their environment. Put them in a village, and they begin to behave like villagers.

Jason Duckworth tells me that until he talked to prospective buyers he expected they would be interested mainly in the houses, but he found that people understood—and liked—the idea of neighborliness. New Daleville's slow start underlines the impor-tance of demonstrating this idea as early as possible. In its next project, Arcadia will make sure that the model homes showcase community features such as ease of walking and common greens.

Developers learn on the job, and Jason has a long mental checklist. He agrees with Comitta that the landscaping can be improved. There needs to be more variety in the sizes and types of houses. He is also dissatisfied with some of the engineering details. The utility company had agreed to locate its equipment in the back lanes, but for lots accessed from the street, it has placed transform-ers and junction boxes next to the sidewalk, which is decidedly unattractive. He believes they can do better in the future.

For developers there's always the next project. Jason, who is busy with several new subdivisions, is spending about half his time working on Arcadia's large new development, Bryn Eyre. The permitting process has been remarkably swift, and construction is expected to begin in two years. The weakening housing market has not affected Arcadia's plans. "We're not pulling back at all," he says. While an economic slowdown may last a year or two, developers' time horizons for getting projects permitted are so much longer that they can't afford to wait. In addition, a weak market creates new opportunities for land acquisition, as cancelled projects come up for sale. Jason and Christy are looking at more than twenty potential sites. As Joe Duckworth told my students, downturns hurt some developers and help others.

Successful developers are not the ones who make the most money in the good times, they're the ones who survive the inevitable bad ones. *All* developers get caught unawares when the market adjusts, but Joe has been careful not to take on too much debt. He has either long-term options with land sellers, or signed contracts with builders. The only project in which he actually owned undeveloped land at the time of the slowdown was Bryn Eyre, but since that development will take twenty years to complete, he's planned for at least two market corrections. So he is taking the current downturn in stride.

For Arcadia, New Daleville has been a roller-coaster ride. The earliest economic projections had shown a modest return on investment; after the builders' auction, lot prices almost doubled; now, thanks to the delays and the weaker market, they are back to where they'd been at the beginning. "It's not as good as I hoped it would be," Duckworth says, "but it's not a tragedy." Nevertheless, the experience has taught him a lesson: neotraditional developments far out in a rural area are risky. The township wanted something different, but the market wasn't quite ready for it. Developers tread a delicate path. They are agents of change, operating between the regulations—and desires—of local jurisdictions and the demands of the marketplace, and they must satisfy both. That isn't always easy, and it's rarely popular.

Four and a half years have passed since Tom Comitta convinced Londonderry Township to try a new sort of development. When I ask Tim Cassidy whether the experiment has converted his neighbors to the neotraditional approach, he says no. So far, most people continue to believe that higher density is not right for a rural community. The planning commission is currently reviewing no less than ten new subdivisions, but none resembles New Daleville; they don't have community water or sewage systems, and lots will be half an acre or larger. That means more scatteration and more exurban growth.

At first glance New Daleville seems to be a case of a strong-willed board of supervisors, supported by a vocal minority on the planning commission, and a persistent developer, pushing an unpopular project through the process. Hardly a propitious model for the future. But the history of development is laden with similar examples. Many fail but some succeed. In any case, change takes time. New Daleville is still not finished. As subdivisions continue to be built in Londonderry—and there's no doubt that they will be—public attitudes may change. Ten years from now, the small lots, the narrow streets, the public park, and the compact cluster of houses on Dr. Wrigley's cornfield will all make sense.

October 2006

A NEW AFTERWORD

New Daleville, November 2007

I spent the earlier part of summer 2007 visiting universities, museums, libraries, and bookstores, lecturing about real estate development and talking about *Last Harvest*. Explaining the complex process of how we build new communities proved more difficult than I expected. The questions people asked revealed that Joe Duckworth was right: nobody trusts developers. Everyone had a pet peeve: the gated community where open fields used to be, the new condominium tower in an old city neighborhood, the sprawling McMansion next door. The elected officials who approved the project, the planners who wrote the zoning ordinance, and the architects who designed the building, all shared in the responsibility, but for the public, the developer was to blame.

Another thing that people didn't seem to understand was that the developer doesn't really control the process: a community can accept or reject a plan; construction costs go up and down; buyers are unpredictable. As my friend and Wharton School colleague Peter Linneman put it, "Good stuff or bad stuff can happen." The worst stuff for a developer is a sudden slowdown in the housing market, I told my audiences, for although market corrections are as regular as clockwork—roughly once a decade—their exact timing is unknowable and the consequences can be grave. After a lecture at Seattle's Town Hall, an obviously incredulous man asked, "But surely developers always make money?" The truth is, no, they do not. Six months ago, Arcadia was obliged to further reduce the price of New Daleville lots to the builder from $105,000 to $90,000. The original bid accepted by Arcadia was $125,000, and the new price was actually less than the cost of acquiring and improving the land. "Thanks to the fact that our prices were initially higher, we're breaking even," Jason Duckworth told me. "The hope is that we can stay alive during this downturn, and recover our costs on the back end." At the present sales rate it would take two or three years to sell the remaining seventy-five lots. Jason hopes that a year from now, prices will have risen, and the final sales—the back end—will make up for current losses. On the other hand, if lot prices remain the same there will be no profit at all, and if they fall, say to $75,000, or if the rate of sales slows down, Arcadia stands to lose a million dollars or more.

Losing money is a distinct possibility since it is unclear when the housing market will recover. According to housing economists, the average downturn lasts less than two years, but this one has gone on for two and a half and shows no sign of ending. The decline in the housing market, which started in July 2005 (when houses went on sale at New Daleville), was aggravated by a crisis in the subprime mortgage market, which lends to home buyers whose income is too low for conventional mortgages. A flurry of foreclosures, combined with overbuilding and the withdrawal of nervous investors and second-home buyers from the market, caused house prices to plummet. Since people don't buy new houses if they can't sell their old ones, the market for new homes effectively stalled. A residential-real-estate-developer friend pithily described the current situation: "When the going was good, we all jumped into the water. Now they're draining the pool, and we're going to find out who's wearing a bathing suit."

The parent company of the two home builders at New

Daleville had a bathing suit, though it was skimpy—NVR's stock value, like that of all the national home builders, had dropped drastically as the market slowed. To reduce overheads at New Daleville, the company withdrew NV Homes from the project, which eliminated one model home and the accompanying sales force, leaving Ryan as the sole builder. House prices were cut further, and in mid-2007 started at $249,000—instead of the previous $275,000, and the original $339,000. The cuts had the desired effect: during summer 2007 Ryan sold between two and three houses per month, more than any other project in southern Chester County.

The result of the sales activity was apparent on the brisk late-November day I walked around New Daleville. Several houses were in various stages of construction, stacks of prefabricated wooden panels awaited assembly, and new excavations stood ready to receive basement walls. There were now enough finished houses—more than forty—that the village atmosphere Joe Duckworth had promised the township was becoming evident. The boulevard was almost completely lined with houses, standing neatly behind concrete sidewalks and white picket fences. The newly planted trees were still small, but backyards were sprouting decks and swing sets, and the community was acquiring a lived-in look. A few front doors had harvest decorations left over from Thanksgiving, and people were starting to put up Christmas wreaths. Several homeowners waved hello as I walked by.

"Everyone who moves in here is attracted by the feeling of neighborliness," Meghan Andress told me when I stopped at her house. She said that last year the residents held a Memorial Day picnic and a community cookout. "We have a mix of people; professionals, a schoolteacher, an FBI agent. They seem to come from all over, Delaware, New Jersey, there is even someone who commutes to work in Philadelphia." She added that a couple of families who had been there less than a year had been transferred to jobs elsewhere and so far had been unable to sell their houses.

Meghan and Scott were unhappy that the value of their home had dropped, but since they were not planning to move in the near future, they were counting on prices rebounding in the long run.

There are many imponderables in real estate development. If the Andresses had waited longer, they would have paid less for their house. If Arcadia's plan for New Daleville had not been so complicated, the project might have been completed more quickly and avoided the market slowdown. It could be argued that the Duckworths paid a serious penalty for trying something new. On the other hand, perhaps not. On my way home, I visited another new subdivision. Now that big expensive houses on large lots were no longer selling, the project was not doing well. A handful of finished houses was surrounded by scores of forlorn LOT AVAILABLE signs, the brand-new streets leading into empty fields produced a derelict air of abandonment. That could have been New Daleville's fate. But thanks to its compact layout, neighborly features, and attractive houses—and Ryan's and Arcadia's willingness to cut prices—the project had persevered. All the time and effort spent on improving the design to make a better community also proved to be an advantage in a difficult economic situation. Good stuff or bad stuff can happen.

December 2007

ACKNOWLEDGMENTS

When I met Joe Duckworth, fifteen years ago during a Harvard summer workshop on house design, I was the teacher and he was the student. During the protracted writing of this book, these roles were reversed. I'm extremely grateful for our frank conversations, as well as for the open access that he provided to the Arcadia Land Company in general and the New Daleville project in particular. Thanks also to Jason Duckworth for his openhanded cooperation, his intelligent observations, and his enthusiastic and helpful assistance throughout. Dave Della Porta and Jim Weidner were patient with my questions and explained technical details. I would also like to acknowledge the help of other members of the New Daleville team: Bob Heuser, Jeff Miller, Simi Baer Kaplin, and Mike DiGeronimo. They were all unfailingly liberal with their time.

Tom Comitta helped me to see New Daleville from the local perspective and also steered me through the zoning literature. Tim Cassidy candidly explained the politics of Londonderry Township and gave me the benefit of his lifelong association with the place. Thanks to Charlotte Wrigley for her reminiscences of the old family farm. Thanks also to Meghan and Scott Andress for generously sharing their time and experiences. I hope they enjoy their new home.

Successful developers are realistic dreamers. Several provided me with valuable professional insights, in particular Robert Davis and Vince Graham, whom I have known and admired for many years. I have written extensively about home building and development, and my conversations with Tom Natelli, Dwight Schar, Robert I. Toll, David Weekley, and the late Charles E. Fraser also informed this book. Thanks, also, to Dick Dilsheimer and Brad Haber. At NVR I want to thank Rick King, Mike Linthicum, Carmela Bond, Greg Norbeck, and especially Kristi Oliveira, for taking the trouble to respond to my too-frequent e-mail questions and telephone calls about their business. Thanks to Amber Britton of Meredith Pepper Inc. for her comments about home decoration.

My appreciation to Peter Linneman, my friend and coeditor at the *Wharton Real Estate Review,* in which a number of articles—by myself and others—prepared the way for this book. Thanks, too, to my Wharton School colleagues Joe Gyourko, Georgette Phillips, and Todd Sinai. At the University of Pennsylvania School of Design, David De Long shared his thoughts about Broadacre City over several agreeable lunches, John Keene explained legal issues in land development, and Jonathan Barnett recalled the early days of New Urbanism. Several architects who are active in neotraditional planning provided helpful observations and recollections, in particular Robert A. M. Stern, Jaquelin T. Robertson, John Massengale, Melanie Taylor, and not least, my friend Andrés Duany. My appreciation to my capable research assistants, Acalya Kiyak, Fernando Moreira, and Ann Lutun.

This was not an easy book to write, and it got finished thanks in large part to four people: my wife, Shirley Hallam, who pushed me to write about this subject and read the manuscript out loud for me, not once but twice; my longtime editor, Nan Graham, who questioned at all the right moments and showed me when too much was too much; my patient publisher, Susan Moldow, who put up with numerous delays and made many useful suggestions; and my steadfast agent, Andrew Wylie.

The Icehouse
Chestnut Hill, Philadelphia
May 2002–October 2006

NOTES

Prologue

1. See David R. Contosta, *Suburb in the City: Chestnut Hill, Philadelphia, 1850–1990* (Columbus: Ohio State University Press, 1992), 83–90.
2. John Stilgoe, *Borderland: Origins of the American Suburb, 1820–1939* (New Haven: Yale University Press, 1988), 308.

Chapter 2: Seaside

1. Robert A. M. Stern, "The Suburban Alternative for the 'Middle City,'" *Architectural Record* 164, no. 2 (August 1978): 96–98.
2. Robert A. M. Stern and John M. Massengale, *The Anglo-American Suburb,* Architectural Design Profile (London: Architectural Design, 1981). Long out of print, this book remains the most comprehensive source of design information about early garden suburbs.
3. See Witold Rybczynski, "Bauhaus Blunders," *Public Interest* no. 113 (Fall 1993): 82–90.
4. Tom Wolfe, "Introduction," *VIA* 4 (1980): 9.
5. I wrote about Seaside in September 24, 1989, in *The New York Times* ("Architects Must Learn to Listen to the Melody"). My first visit was a month later, as architecture critic for *Wigwag* (see "Our Town" in the March 1990 issue).
6. Quoted by Lewis Mumford, *The City in History: Its Origins, Its Transformations, and Its Prospects* (New York: Harcourt, Brace & World, 1961), 332.
7. See Philip Langdon, *A Better Place to Live: Reshaping the American Suburb* (Amherst: University of Massachusetts Press, 1994), 119.
8. See Witold Rybczynski, "Tomorrowland," *New Yorker* (July 22, 1996): 36–39.
9. Andrés Duany and Elizabeth Plater-Zyberk, "The Second Coming of the American Small Town," *Wilson Quarterly* 16 (Winter 1992): 21.
10. Ibid., 22.
11. Ibid., 40.

Chapter 3: Epiphanies

1. See Witold Rybczynski, "The Art of the New Urbanist Deal," *Wharton Real Estate Review* 6, no. 2 (Fall 2002): 56–64.
2. See Frank Jackson, *Sir Raymond Unwin: Architect, Planner and Visionary* (London: A. Zwemmer, 1985).
3. Raymond Unwin, *Town Planning in Practice: An Introduction to the Art of Designing Cities and Suburbs* (London: T. Fisher Unwin, 1909; New York: Princeton Architectural Press, 1994).

Chapter 4: Last Harvest

1. Alecia Swasy, "America's 20 Hottest White-Collar Addresses," *Wall Street Journal* (March 8, 1994): B1.
2. Amy Donohue, "Is Chester County the New Main Line?" *Philadelphia* (April 2003).
3. Dan Rose, *Patterns of American Culture: Ethnography and Estrangement* (Philadelphia: University of Pennsylvania Press, 1989), 49.
4. Nancy L. Mohr, *The Lady Blows a Horn* (Boyertown, Pa.: Horse Country Press, 1997), 114.
5. Tom Lea, *The King Ranch* (Kingsville, Tex.: King Ranch, 1957), 767.
6. Mohr, *Lady Blows a Horn*, 92.
7. See Daniel R. Mandelker, *Land Use Law* (Newark, N.J.: Matthew Bender & Co., 2003), 1–4.
8. Marc A. Weiss, *The Rise of the Community Builders: The American Real Estate Industry and Urban Land Planning* (New York: Columbia University Press, 1987), 81.
9. Ibid., 85; Yale Rabin, "Expulsive Zoning: The Inequitable Legacy of *Euclid*," in *Zoning and the American Dream: Promises Still to Keep*, Charles M. Haar and Jerold S. Kayden, eds. (Chicago: Planners Press, 1989), 103–7.
10. William A. Fischel, *The Economics of Zoning Laws: A Property Rights Approach to American Land Use Controls* (Baltimore: Johns Hopkins University Press, 1985), 29.
11. *National Land and Investment Co. v. Easttown Twp. Board of Adjustment*, 419 Pa. 504, 215, A.2d 597 (1965).
12. *Concord Township* Appeal, 439 Pa. 466; 268 A.2d 765 (1970).
13. *Girsh* Appeal, 437 Pa. 237, 263 A.2d 395 (1970).

Chapter 5: Life, Liberty, and the Pursuit of Real Estate

1. Quoted by John W. Reps, *The Making of Urban America: A History of City Planning in the United States* (Princeton: Princeton University Press, 1965), 167.
2. Ibid., 166.
3. Howard Malcolm Jenkins, *The Family of William Penn, Founder of Pennsylvania, Ancestors and Descendents* (Philadelphia: Howard Jenkins, 1899), 219–20, 246.
4. J. Smith Futhey and Gilbert Cope, *History of Chester County, Pennsylvania, with Geological and Biographical Sketches* (Philadelphia: Louis H. Everts, 1881), 174.
5. Jenkins, *Family of William Penn*, 65.
6. Quoted by A. M. Sakolski, *The Great American Land Bubble: The Amazing Story of Land-Grabbing, Speculations, and Booms from Colonial Days to the Present Time* (New York: Harper & Brothers, 1932), 7.
7. Ibid., 11.
8. Robert D. Arbuckle, *Pennsylvania Speculator and Patriot: The Entrepreneurial John Nicholson, 1757–1800* (University Park: Pennsylvania State University Press, 1975), 5–7.
9. Robert D. Arbuckle, "John Nicholson and Land as a Lure in the Infant Nation," *Pennsylvania Heritage* 9, no. 2 (Spring 1983): 9.
10. Sakolski, *Great American Land Bubble*, 156–57.
11. Quoted by William Graham Sumner, *The Financier and the Finances of the American Revolution*, vol. 2 (1891; repr., New York: Augustus M. Kelley, 1968), 241.
12. Ibid., 265.
13. Quoted by Arbuckle, *Pennsylvania Speculator*, 197.
14. Ibid., 204.

Chapter 6: Joe's Deal

1. There is a distinct correlation between pro- and anti-growth areas and the division of America in red and blue voting blocs. See Joel Kotkin, "Suburban Tide," *Blueprint* (March 15, 2005).
2. Hazel A. Morrow-Jones et al. "Consumer Preference for Neotraditional Neighborhood Characteristics," *Housing Policy Debate* 15, no. 1 (2004): 195, 186.
3. Ibid., 196.

NOTES

Chapter 7: On the Bus

1. *The Lexicon of the New Urbanism* (Miami: Duany Plater-Zyberk & Co., 1999), F-3.1.
2. See Mark J. Eppli and Charles C. Tu, *Valuing the New Urbanism: The Impact of the New Urbanism on Prices of Single-Family Houses* (Washington, D.C.: Urban Land Institute, 1999).

Chapter 8: Meetings

1. Robert Fishman, *Bourgeois Utopias: The Rise and Fall of Suburbia* (New York: Basic Books, 1987), 67.

Chapter 9: Scatteration

1. Anthony Downs, "Some Realities About Sprawl and Urban Decline," *Housing Policy Debate* 10, no. 4 (1999): 961.
2. Stephen Malpezzi, "Estimates of the Measurement and Determinants of Urban Sprawl in U.S. Metropolitan Areas" (unpublished manuscript, Center for Urban Land Economic Research, University of Wisconsin, Madison, 1999), 23–27.
3. See William A. Fischel, *The Economics of Zoning Laws: A Property Rights Approach to American Land Use Controls* (Baltimore: Johns Hopkins University Press, 1985).
4. Gregg Easterbrook, "Suburban Myth," *New Republic* (March 15, 1999): 19.
5. John Tierney, "The Autonomist Manifesto (Or, How I Learned to Stop Worrying and Love the Road)," *New York Times Magazine* (September 26, 2004): 61.
6. Reid H. Ewing, "Characteristics, Causes, and Effects of Sprawl: A Literature Review," *Environmental and Urban Issues* 15, no. 2 (January 1988): 1.
7. Peter Gordon and Harry Richardson, Letters to the editors, *Journal of the American Planning Association* 63, no. 2 (Spring 1997): 27.
8. Randal O'Toole, *The Vanishing Automobile and Other Urban Myths* (Brandon, Ore.: Thoreau Institute, 2001), 392.
9. Witold Rybczynski, "Measuring Sprawl," *Wharton Real Estate Review* 6, no. 1 (Spring 2002): 101–2.
10. Jerry Adler, "Bye, Bye, Suburban Dream," *Newsweek* (May 15, 1995): 45.
11. Rybczynski, "Measuring Sprawl," 101–2.
12. Peter Gordon and Harry W. Richardson, "Are Compact Cities a Desirable Goal?" *Journal of the American Planning Association* 63, no. 1 (Winter 1997): 103.

13. Downs, "Some Realities," 955.
14. Development Panel, Zell-Lurie Real Estate Center Members Meeting, Philadelphia, October 21, 2003.
15. See, for example, F. Kaid Benfield et al., *Once There Were Greenfields: How Urban Sprawl Is Undermining America's Environment, Economy, and Social Fabric* (New York: Natural Resources Defense Council, 1999).
16. George Galster et al., "Wrestling Sprawl to the Ground: Defining and Measuring an Elusive Concept," *Housing Policy Debate* 12, no. 4 (2001): 681–82.
17. Lewis Mumford, *The City in History: Its Origins, Its Transformations, and Its Prospects* (New York: Harcourt, Brace & World, 1961), 511.
18. Robert O. Harvey and W.A.V. Clark, "The Nature and Economics of Urban Sprawl," *Land Economics* 41, no. 1 (February 1965): 1–9.
19. Edward P. Eichler and Marshall Kaplan, *The Community Builders* (Berkeley: University of California Press, 1967), 169.
20. Real Estate Research Corporation, *The Costs of Sprawl: Environmental and Economic Costs of Alternative Residential Development Patterns at the Urban Fringe,* vol. 1, *Detailed Cost Analysis* (Washington, D.C.: U.S. Government Printing Office, 1974).
21. See Robert W. Burchell et al., *The Costs of Sprawl—Revisited* (Washington, D.C.: National Academy Press, 1998), i; Benfield et al., *Once There Were Greenfields,* 96–97.
22. Burchell et al., *Costs of Sprawl,* 26.
23. William Schneider, "The Suburban Century Begins," *Atlantic Monthly* (July 1992): 33.
24. John Nolen, *New Towns for Old* (1927; repr., London: Routledge/Thoemmes Press, 2001), 119.
25. Easterbrook, "Suburban Myth," 19.
26. Karen A. Danielson et al., "Retracting Suburbia: Smart Growth and the Future of Housing," *Housing Policy Debate* 10, no. 3 (1999): 513–15.
27. Anthony Downs, "What Does 'Smart Growth' Really Mean?" *Planning* 67, no. 4 (April 2001): 23, 25.

Chapter 10: More Meetings

1. James R. Hagerty, "Home Construction Continues at Robust Pace," *Wall Street Journal* (January 20, 2005): D2.

Chapter 12: On the Way to Exurbia

1. The term was coined by Auguste C. Spectorsky in *The Exurbanites* (Philadelphia: J. B. Lippincott, 1955).
2. David Brooks, "Take a Ride to Exurbia," *New York Times* (November 9, 2004): A23.
3. H. G. Wells, *Anticipations: Of the Reaction of Mechanical and Scientific Progress upon Human Life and Thought* (Leipzig: Bernhard Tauchnitz, 1902).
4. Ibid., 47.
5. David G. De Long, "Frank Lloyd Wright and the Evolution of the Living City," in *Frank Lloyd Wright and the Living City*, David G. De Long, ed. (Milan: Skira editore, 1998), 20–24.
6. David G. De Long, "Designs for an American Landscape, 1922–1932," in *Frank Lloyd Wright: Designs for an American Landscape, 1922–1932*, David G. De Long, ed. (New York: Harry N. Abrams, 1996), 20.
7. A streetcar line was planned for Palos Verdes Estates, but by the time that construction began, the automobile was ascendant and the line was never built. For background on Palos Verdes, see Alexander Garvin, *The American City: What Works, What Doesn't* (New York: McGraw-Hill, 1996), 331–35.
8. Only a few drawings of the Doheny project have survived but no site plan. The project is reconstructed in De Long, "Designs for an American Landscape," 19–30.
9. Ibid., 20.
10. This was three years before he actually visited New York City. Le Corbusier, "A Noted Architect Dissects Our Cities," *New York Times Magazine* (January 3, 1932): 10–19.
11. Ibid., 11.
12. The famous International Style exhibition at the Museum of Modern Art in New York opened in February 1932.
13. Frank Lloyd Wright, "'Broadacre City': An Architect's Vision," *New York Times Magazine* (March 20, 1932): 8–9.
14. Ibid., 8.
15. Frank Lloyd Wright, *The Disappearing City* (New York: William Farquhar Payson, 1932), 44.
16. Quoted by John Sergeant, *Frank Lloyd Wright's Usonian Houses: The Case for Organic Architecture* (New York: Whitney Library of Design, 1975), 134.
17. Frank Lloyd Wright, *When Democracy Builds* (Chicago: University of Chicago Press, 1945), 38.
18. Sergeant, *Frank Lloyd Wright's Usonian Houses,* 134.
19. Frank Lloyd Wright, *The Living City* (New York: Horizon Press, 1958), 230.
20. Wright, "'Broadacre City,'" 9.

21. For example, De Long, "Frank Lloyd Wright and the Evolution of the Living City," 28.
22. Wright, *When Democracy Builds,* opp. 55.
23. George Collins, "Broadacre City: Wright's Utopia Reconsidered," in *Four Great Makers of Modern Architecture: Gropius, Le Corbusier, Mies van der Rohe, Wright* (New York: Trustees of Columbia University, 1963), 67; Neil Levine, *The Architecture of Frank Lloyd Wright* (Princeton: Princeton University Press, 1996), 220; Ada Louise Huxtable, *Frank Lloyd Wright* (New York: Viking Penguin, 2004), 201.
24. Brendan Gill, *Many Masks: A Life of Frank Lloyd Wright* (New York: G.P. Putnam's Sons, 1987), 337–38.
25. Lewis Mumford, *The City in History: Its Origins, Its Transformations, and Its Prospects* (New York: Harcourt, Brace & World, 1961), 633.
26. Joel Garreau, *Edge City: Life on the New Frontier* (New York: Doubleday, 1991), 10.
27. Frank Lloyd Wright, *The Natural House* (New York: Horizon Press, 1954), 134.

Chapter 13: Design Matters

1. Raymond Unwin, *Town Planning in Practice: An Introduction to the Art of Designing Cities and Suburbs* (London: T. Fisher Unwin, 1909; New York: Princeton Architectural Press, 1994), 367.
2. Ibid.
3. Susan L. Kaus, *A Modern Arcadia: Frederick Law Olmsted Jr. and the Plan for Forest Hills Gardens* (Amherst: University of Massachusetts Press, 2002), 91.
4. *Protective Restrictions Palos Verdes Estates* (Palos Verdes Estates: Los Angeles, 1923), 1.
5. Anton C. Nelessen, *Visions for a New American Dream: Process, Principles, and an Ordinance to Plan and Design Small Communities* (Chicago: Planners Press, 1994), 81–102.
6. Ibid., 91.
7. *Pennsylvania Municipalities Planning Code,* Article VII-A, 68.

Chapter 14: Locked In

1. Clifford J. Treese, *Community Associations Factbook* (Alexandria, Va.: Community Associations Institute, 1999), 19.
2. "Declaration of Covenants, Easements, Conditions and Restrictions of New Daleville, A Planned Community," New Daleville Associates, L.P., January 16, 2004, 41–45.
3. American Housing Survey, 2001; Community Associations Institute, 2003.

4. Evan McKenzie, *Privatopia: Homeowner Associations and the Rise of Residential Private Government* (New Haven: Yale University Press, 1994), 122–49.
5. Richard Briffault, "A Government for Our Time: Business Improvement Districts and Urban Governance," *Columbia Law Review* 365 (March 1999).
6. *National Survey of Community Associations Homeowner Satisfaction* (Alexandria, Va.: Community Associations Institute, 1999), 6, 5.
7. Quoted by John Tierney in "The Mansion Wars," *New York Times* (November 15, 2005): A27.

Chapter 15: House and Home

1. National Association of Home Builders, 2003.
2. I have written about the Dutch episode in *Home: A Short History of an Idea* (New York: Viking Penguin, 1986), 51–75.
3. Stefan Muthesius, *The English Terraced House* (New Haven: Yale University Press, 1982), 1.
4. D. Benjamin Barros, "Home as a Legal Concept," *Santa Clara Law Review* 46, no. 2 (2006): 260.
5. *Housing Statistics in the European Union 2002* (Delft: OTB Research Institute for Housing, Urban and Mobility Studies, 2000). Two-thirds of the Norwegian housing stock is houses, although this is slightly overstated since Norwegian statistics lump together one- and two-family houses. *Housing Statistics* (Oslo: Norwegian State Housing Bank, July 2003).
6. Houses are 90 percent of the housing stock in Ireland, 80 percent in Australia, and 70 percent in Canada. *Australian State of the Environment Report, 2001;* Canada Mortgage and Housing Corporation, *2004 Canadian Housing Observer.*
7. Sixty percent of urban dwellings in the United States are single-family houses. *National Housing Survey,* 2001.
8. Werner Blaser, *Courtyard Houses in China: Tradition and Present* (Basel: Birkhäuser Verlag, 1979), 5.
9. *2001 English House Condition Survey* (London, July 2003).
10. *National Housing Survey,* 2001.
11. Robert Bruegmann, *Sprawl: A Compact History* (Chicago: University of Chicago Press, 2005), 201.

Chapter 16: Generic Traditional

1. Andrés Duany et al., *Suburban Nation: The Rise of Sprawl and the Decline of the American Dream* (New York: Northpoint Press, 2000), 208.
2. Kurt Andersen, "Is Seaside Too Good to Be True?" in *Seaside: Making a Town in America,* David Mohney and Keller Easterling, eds. (New York: Princeton Architectural Press, 1991), 46.

3. Alan Gowans, *The Comfortable House: North American Suburban Architecture* (Cambridge, Mass.: MIT Press, 1986), 141.
4. David Brooks, *On Paradise Drive: How We Live Now (And Always Have) in the Future Tense* (New York: Simon & Schuster, 2004), 50–51.
5. Beth Dunlop, *Building a Dream: The Art of Disney Architecture* (New York: Harry N. Abrams, 1996), 28.
6. Witold Rybczynski, "Tomorrowland," *New Yorker* (July 22, 1996): 36–39.
7. Robert A. M. Stern, "Jumping-Off Points," *ANY* 1 (July–August 1993): 45.
8. *Celebration Pattern Book* (Celebration, Fla.: Walt Disney Company, March 1995), A3.

Chapter 17: The Dream

1. Quoted in Allen Sinclair Will, "America, Nation of Dreamers," *New York Times* (October 4, 1931): 61.
2. Marc A. Weiss, *The Rise of the Community Builders: The American Real Estate Industry and Urban Land Planning* (New York: Columbia University Press, 1987), 1.
3. Kenneth T. Jackson, *Crabgrass Frontier: The Suburbanization of America* (New York: Oxford University Press, 1985), 234–36.
4. Michel T. Kaufman, "Tough Times for Mr. Levittown," *New York Times Magazine* (September 24, 1989): 44.
5. Barbara M. Kelly, *Expanding the Dream: Building and Rebuilding Levittown* (Albany: State University of New York Press, 1993), 26.
6. Ibid., 189n15.
7. Michele Ingrassia, "The House That Levitt Built," *Newsday.com* (2005).
8. Frank Lloyd Wright, *The Natural House* (New York: Horizon Press, 1954), 68.
9. John Sergeant, *Frank Lloyd Wright's Usonian Houses: The Case for Organic Architecture* (New York: Whitney Library of Design, 1975), 16.
10. William Allin Storrer, *The Frank Lloyd Wright Companion* (Chicago: University of Chicago Press, 1993), 252.
11. Ron Rosenbaum, "The House That Levitt Built," *Esquire* (December 1983): 385.
12. Kelly, *Expanding the Dream*, 35.
13. Alexander Garvin, *The American City: What Works, What Doesn't* (New York: McGraw-Hill, 1996), 337.
14. Herbert J. Gans, *The Levittowners: Ways of Life and Politics in a New Suburban Community* (New York: Pantheon Books, 1967), xvi.
15. Ibid., 293.
16. Ibid., 413.
17. Ibid., 432.
18. "New Towns," *Architectural Forum* (November 1951): 138.
19. "Levitts Plan Many 'Extras' in Homes for Defense Workers in Bucks County," *New York Times* (November 4, 1951): 1.

20. Ingrassia, "House That Levitt Built."
21. "New Levitt Houses Break All Records," *House + Home* (February 1952): 98.

Chapter 18: Builders

1. Jaquelin T. Robertson, "The House as the City," in *New Classicism,* Andreas Papadakis and Harriet Watson, eds. (London: Academy Editions, 1990), 234.
2. "1992 Census of Construction Industries" (U.S. Department of Commerce, Bureau of the Census, June 1995).
3. *Builder* (May 2005).
4. The top ten nationals in 2005 in terms of annual housing starts were D. R. Horton, Pulte Homes, Lennar Corporation, Centex Corporation, KB Home, Beazer Homes USA, Hovnanian Enterprises, Ryland Group, MDC Holdings, and NVR (*Builder,* May 2006).
5. This example is based on Ryan Homes's Avalon model in 2005. See ryan-homes.com.
6. "J. D. Power and Associates Reports: Pulte Homes Receives Platinum Award for Excellence on Home Building and Ranks Highest in Customer Satisfaction in Fourteen U.S. Markets," J. D. Power and Associates Press Release, September 15, 2004.
7. "Region's Home Builders Fare Poorly in Survey," *Philadelphia Inquirer* (September 16, 2004).

Chapter 19: A Compromise

1. Timothy J. Cassidy, "Critical Regionalism—A Reflective Perspective from the Brandywine Valley" (Ph.D. diss., Texas A&M University, May 2000), 217.

Chapter 20: Trade-offs

1. See Witold Rybczynski, "Dream Town, USA," posted on Slate.com, February 9, 2005.

Chapter 21: Mike and Mike

1. Edward P. Eichler and Marshall Kaplan, *The Community Builders* (Berkeley: University of California Press, 1967), 146.
2. Alan Gowans, *The Comfortable House: North American Suburban Architecture, 1890–1930* (Cambridge, Mass.: MIT Press, 1986), 59–63.

3. Alfred Bank and Harold Sandbank, *A History of Prefabrication* (New York: Arno Press, 1972), 57.

Chapter 22: Ranchers, Picture Windows, and Morning Rooms

1. Alfred Bank and Harold Sandbank, *A History of Prefabrication* (New York: Arno Press, 1972), 29–31.
2. Ibid., 44–52.
3. Gilbert Herbert, *The Packaged House: Dream and Reality* (Haifa: Technion, Documentation Unit of Architecture, 1981).
4. *Housing Facts, Figures, and Trends 2004* (Washington, D.C.: National Association of Home Builders, January 2005).
5. Charles Moore, Gerald Allen, and Donlyn Lyndon, *The Place of Houses* (New York: Holt, Rinehart and Winston, 1974), 107.
6. *Housing Facts.*
7. Ibid., 11.
8. Ibid., 10.
9. Ibid.
10. *NVR Annual Report 2004*, 7.

Chapter 24: The Market Rules

1. *The Linneman Letter* (Fall 2005), 26.
2. Most prominently, Robert J. Schiller, "The Bubble's New Home," *Barron's* (June 20, 2005); and Paul Krugman, "The Hissing Sound," *New York Times* (August 8, 2005): A15. The counterargument was presented by Chris Mayer and Todd Sinai, "Bubble Trouble? Not Likely," *Wall Street Journal* (September 19, 2005): A16.
3. Karl E. Case and Robert J. Schiller, "Is There a Housing Bubble?" *Brookings Papers on Economic Activity* (Brookings Institution 2003): 2, 299.
4. Margaret Hwang Smith, Gary Smith, and Chris Thompson, "When Is a Housing Bubble Not a Housing Bubble?" (Unpublished paper, Pomona College, Claremont, Calif., 2005), 3.
5. See Charles Himmelberg, Christopher Mayer, and Todd Sinai, "Assessing High House Prices: Bubbles, Fundamentals, and Misperceptions," *Journal of Economic Perspectives* 19, no. 4 (Fall 2005).
6. National Association of Home Builders (November 17, 2005).
7. *New York Times* (October 4, 2005): A1.
8. *Atlanta Journal-Constitution* (November 2005): 1C.

Chapter 25: Bumps in the Road

1. *Washington Post* (February 28, 2006): D03.
2. James O'Sullivan, quoted in *The New York Times* (March 1, 2006): C3.
3. *Financial Times* (March 26, 2006): 1.

Chapter 29: Moving Day

1. Allan Greenberg, *George Washington Architect* (London: Andreas Papadakis Publisher, 1999), 115.

INDEX

ABOUT THE AUTHOR

Witold Rybczynski has written about architecture for *The New York Times, Time, The Atlantic, The New Yorker,* and *Slate.* He is the author of the critically acclaimed book *Home* and the award-winning *A Clearing in the Distance.* The recipient of the National Building Museum's 2007 Vincent Scully Prize, he lives with his wife in Philadelphia, where he teaches at the University of Pennsylvania's School of Design.